# BRIDGES EXPLAINED
## Viaducts • Aqueducts

––––––––– TREVOR YORKE –––––––––

COUNTRYSIDE BOOKS
NEWBURY, BERKSHIRE

First published 2008
© Trevor Yorke, 2008
Reprinted 2025

All rights reserved. No reproduction
permitted without the prior permission
of the publisher:

COUNTRYSIDE BOOKS
3 Catherine Road
Newbury, Berkshire

To view our complete range of books,
please visit us at
www.countrysidebooks.co.uk

ISBN 978 1 84674 079 4

Drawings and photographs by the author
except for Fig.1.2: Julie Pointon;
Figs. 6.4, 6.15, 8.12, 8.13, 8.14, 8.17, 10.7,
10.12, 10.13, 11.6 and 11.14 (bottom): Stan Yorke;
Fig. 9.19 (top): Eric Brady

*Thanks are also due to the Property Department
of the Manchester Ship Canal Company for all their
help with information on the Barton Swing Aqueduct*

Produced through MRM Associates Ltd., Reading
Typeset by Jean Cussons Typesetting, Diss, Norfolk
Printed by Holywell Press, Oxford

# Contents

Introduction 5

## SECTION I

### Chapter 1

BUILDING BRIDGES

7

## SECTION II

### Chapter 2

ROMAN AND MEDIEVAL ARCHED BRIDGES

21

### Chapter 3

TUDOR AND STUART ARCHED BRIDGES

40

### Chapter 4

GEORGIAN ARCHED BRIDGES

48

### Chapter 5

CAST-IRON, STEEL & CONCRETE ARCHED BRIDGES

61

## SECTION III

### Chapter 6

BEAM, TRUSS AND GIRDER BRIDGES

72

### Chapter 7

CANTILEVER BRIDGES

83

**Chapter 8**
SUSPENSION BRIDGES
89

**Chapter 9**
MOVING BRIDGES
114

## SECTION IV

**Chapter 10**
VIADUCTS
129

**Chapter 11**
AQUEDUCTS
137

## SECTION V

BRIDGES TO VISIT
146

GLOSSARY
156

BIBLIOGRAPHY
158

INDEX
159

# Introduction

Bridges are a unique and vital form of structure. They are key to communication and transport and without them the country would not have developed and could not operate today. Bridges link communities, making parts of a town or city that are separated by roads or rivers for example into one. Many draw in tourists, while their outline can feature as a regional logo or on coins, iconic symbols recognised worldwide.

For all their essential functionality bridges also have an aesthetic and environmental value. They are by their simple geometric form and structural use of material an attractive feature in the landscape. Stripped of decoration the graceful curve of an arch, the fine horizontal line of a beam or the elegant hung cables of a suspension bridge complement the surrounding countryside. They stand as man's imposition on, and conquest of, nature yet at the same time hark back to a prehistoric worship of and respect for the silvery waters they cross.

For me, bridges evoke a wide range of emotions: that very English deep set love of the rustic, the wonder of a spectacular modern structure vanishing into the mist, and the shear terror when crossing anything more than fifty feet below! It was whilst spending hours studying bridges at the drawing board that I started to ask questions about how they were built, what exactly a particular part did, and why they didn't fall down! The answers, however, could be hard to find. Few bridges have visitor centres and books on the subject seem rather sparse despite the public's obvious interest. It was this gradual frustration with the lack of easily accessible and understandable information that I set out to learn for myself how, when, and why bridges were built

The book uses diagrams, drawings, and photographs to explain the basic principles of construction and styles of bridges so that the reader can better appreciate them, recognize how they work, and identify the period from which individual bridges probably date. The book is divided into four sections, the first giving a background to the subject. The second covers arched bridges, the third other forms of bridges, and the fourth viaducts and aqueducts. Finally, there is a list of all the bridges included in the book with a grid reference to accurately locate them, and a glossary of common terms.

Although it is hoped this book will expand your knowledge of the subject it is also intended to inspire you to visit different examples or look again at familiar bridges in a new light.

Trevor Yorke

# THE VARIOUS PARTS THAT MAKE UP A BRIDGE

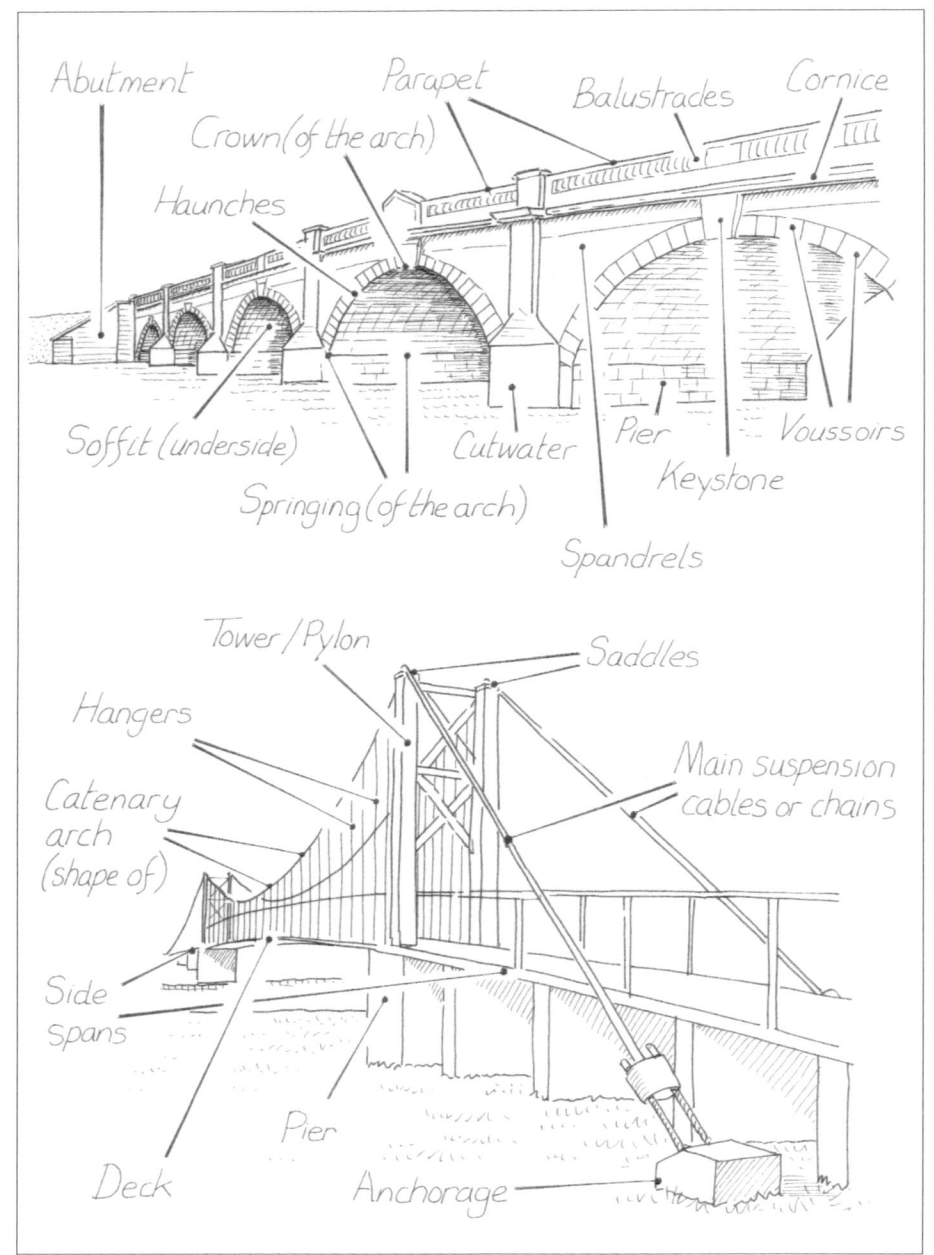

# Chapter 1

# *Building Bridges*

**FIG. 1.1:** *The Tay Bridge Disaster: The famous collapse of what was then the world's longest bridge highlights many of the difficulties which face all bridge builders.*

A ferocious storm had suddenly built up on the east coast of Scotland during the afternoon of Sunday 28th December 1879. All had been calm six hours earlier but now, after dark, gale force winds were rattling down the Forth of Tay and over the city of Dundee. At the railway station staff were growing concerned that the train from Edinburgh due in at 7.15 pm had not arrived, and members of the public were adding to their anxiety by stating that they had seen a flash of fire along the new bridge on the approach to the station. This vast iron girder structure stretched over the estuary for nearly two miles. It had been opened only the previous year to much praise, especially for its designer, Thomas Bouch, who was subsequently knighted by the Queen after she herself had made the crossing.

# Bridges Explained: Viaducts & Aqueducts

In the blackness of this stormy winter's night it was up to the station master and locomotive foreman to investigate. Having confirmed with the signalman that the missing train had entered the southern end of the bridge at 7.14 pm but that communication had since been lost with the signal box at the other end, there was nothing else to do but walk out along the bridge to find out what had happened. They stepped into the gloom and made their way to the middle of the bridge, finding it increasingly hard the further they went to keep their footing in the battering side winds. Then they were suddenly brought to a stop; the bridge in front of them had vanished. The men were standing above a pier with a huge void and the raging sea below. Convinced they could see a red light on the other end of the bridge which might be the missing train, the men were hopeful that it had been pulled up short of the fall, but, with the discovery of baggage washed ashore, it became clear on their return that it was not so.

With daylight the full scale of the disaster unfolded. The whole central section of the bridge, 13 spans, known as the high girders, as they permitted a greater clearance below for the passage of shipping, had simply gone, taking the locomotive, carriages, and 75 passengers and crew with them. Boats had been out in the night but had failed to find any survivors, in part a terrible consequence of the compulsory locking of carriage doors on British railways at the time.

In the aftermath it was the respected engineer Thomas Bouch who took the blame for the disaster. Because the surveys of the sea bed had proved to be inaccurate, he had changed his design from one using brick piers to one with iron columns in order to lighten the load on the foundations. This new structure was tall and narrow, with inadequate cross-bracing to give the structure stability, so that when it was hit by a strong cross wind, as happened on this night, the ironwork failed and the structure toppled into the sea. Bouch, who at the time had already laid the foundation stone for his next project, the Forth Rail Bridge, found his career in ruins and he died later the following year, reportedly a broken man.

FIG. 1.2: THE TAY RAIL BRIDGE, DUNDEE: *In the right of this picture can be seen the remains of the brick and stone piers of the original bridge alongside the replacement bridge of 1887 to the left.*

A new bridge across the Tay was built alongside the old one, this time with a double track, thus creating a wider and substantially stronger structure, which stands to this day. The brick pier bases of the old bridge were retained to act as breakwaters for the new structure upstream of it, and some of the girders were re-used, albeit with additional strengthening. Incredibly, the locomotive which plunged into the sea on that fateful day was salvaged and repaired, and carried on in service for another 25 years, although it was always known to railway men as 'the Diver'!

**ENGINEERING A BRIDGE**
The story of this well-known disaster highlights the difficulties which face all bridge builders and are worth explanation before looking at the different types of structures in the following sections. Whether it is simply a stone or timber beam across a brook or the latest suspension bridge spanning an estuary, the designer has to tackle the same basic problems. Where is it best to build it? What are the forces which will act upon it? What materials are suitable for it? How will it be built?

## Location
The siting of a bridge is one of the most important decisions in the design process. The abutments and piers or towers need to be built upon a firm footing, and so discovering the geology of the land and the river bed is the first step and this will have an effect upon the choice of structure and materials used for the bridge. Although the narrowest part of the river may seem the ideal site, a place where the bedrock or best ground is nearest to the surface is more likely to be chosen. Often this will be where a previous structure and/or ford stood, as this will indicate where the river bed and flow is suitable and will tie in with the existing road network. The ford can also continue in use for heavier vehicles to reduce wear on the new bridge.

The sides of a valley or the embankments on the approach to a bridge, viaduct, or aqueduct must also be carefully studied and engineered. At Ironbridge the notoriously unstable

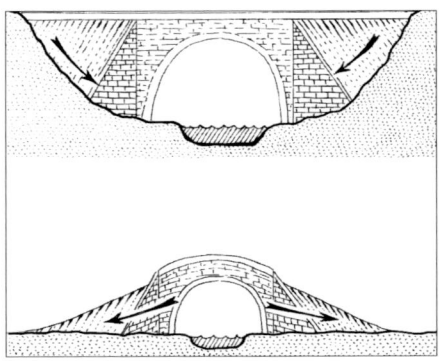

**FIG. 1.3:** *The form of the bridge and its abutments will be shaped by the lie of the land. In a steep valley the abutments must be well secured to stop them slipping down the sides and compressing the bridge. On a flat site where traffic passes beneath it, the bridge must be raised for clearance, creating a humpback bridge which is awkward to cross and needs to be well supported to avoid the arch pushing the abutments outwards.*

# Bridges Explained: Viaducts & Aqueducts

geology of the gorge caused the southern approach to slide and fail, and it had to be replaced only a few decades after the bridge was opened. The embankments leading up to a bridge also exert a lateral force as they slowly settle and slide over time and so

FIG. 1.4: WILLOWFORD BRIDGE, HADRIAN'S WALL, NORTHUMBERLAND: *This view looks down on the remains of the eastern end of the old Roman bridge, with Hadrian's Wall running above it. The river has moved in the 1800 years since the bridge was built and now lies a couple of hundred yards to the west, with fields where it once flowed under the bridge. This demonstrates how far rivers can move in a short geological time.*

abutments which wrap around the exposed end need to be substantial enough and buttressed to resist this gradual spread.

Finding a good footing on land is easier than in water, yet there is still no guarantee that it will be straightforward. Rivers have a habit of changing course over time and what might be dry land today may have once been the river bed or at least be built up of deposits (alluvium) left by centuries of flooding. It was the discovery that there was not adequate bedrock below the projected course of the Tay Bridge which forced Thomas Bouch to make the fateful change to his design that eventually resulted in its failure.

## The Forces of Nature and Man

Once the site for a bridge has been chosen, the designer will have to consider the forces which will act upon it in this location. These come from three sources: the weight of the structure itself, known as the dead load; the pressure weighing down upon it from the crossing traffic, called the live load; and the forces applied by nature, like wind or river flow.

The dead load is usually straightforward to predict and rarely causes failure. The live load can lead to problems from vibrations induced by movement and from unevenly distributed loads. A road or footbridge usually has traffic spread over the deck, but on a railway bridge an immense weight in the form of a train is applied on just one part at a time. Suspension bridges were used sometimes on early

# Building Bridges

**FIG. 1.5 (*right*):** *Drawing showing the forces which the engineer has to consider when designing a bridge.*

**FIG. 1.6 (*below*): OAKLEY BRIDGE, BEDFORDSHIRE:** *The height to which a seemingly tranquil river can rise is frightening. The Severn, powered by masses of melt water from the Welsh mountains, is a well known example, but even rivers rising in relatively flat land, like the Great Ouse in this example, can flood to amazing levels, as testified by the marker which records the events of 1 November 1823, when the level reached above the height of the parapets, completely submerging the structure.*

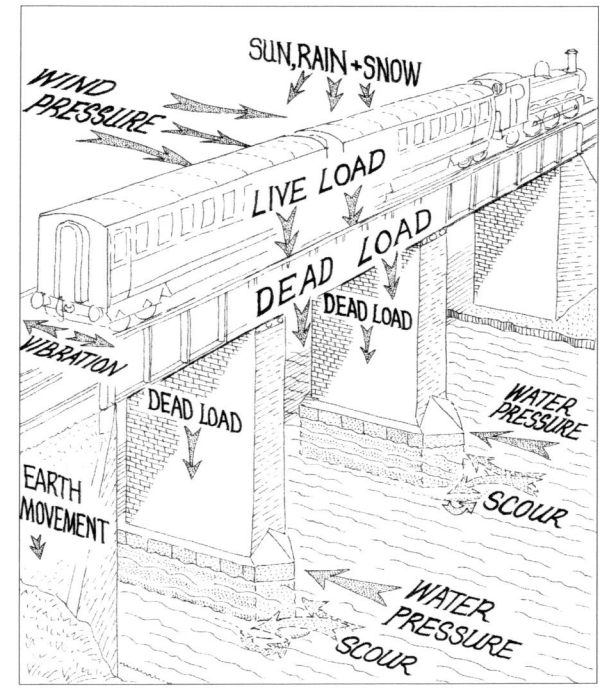

railways in this country, but because this factor caused dangerous movement such structures were usually replaced.

The forces of nature are more problematical than the dead and live loads. The pressure applied by water and wind varies with weather and season, lateral forces usually having the most significant effects. The flow of a river is the greatest problem and so, in order reduce its damaging effects, the water is normally directed round the piers and towers by means of pointed or rounded cutwaters. At times of flood, however, the river rises above these, greatly increasing the pressure upon the flat face of the bridge. The problem is intensified by debris carried down the river; uprooted trees can hit the structure like a battering ram, or become wedged in the gaps, forcing the water to rise still further. Most stone and timber bridges across rivers have been rebuilt at least once because of damage sustained during flooding. Another problem is scouring caused by the water current around the base of the piers, which can undermine them, especially if they have not been properly bedded on a firm footing.

Wind pressure becomes a problem with the longer, lighter and taller structures which began appearing in the late 18th century. It was the principal force in the collapse of the Tay Bridge. Thomas Bouch had believed the bridge had to be strong enough to withstand only a maximum wind pressure of 10 lbs per square foot, but on the night of its collapse the wind force greatly exceeded this figure, and following the subsequent inquiry it was recommended that future bridges should be able to take a pressure of more than five times that allowed for by Bouch.

## Materials

The three forces acting upon a bridge create stresses in the structure which the engineer has to combat by selecting materials with the most suitable properties for his design. There are four types of stresses which are considered in the design of bridges, the first two, and most important, being compression (pushing) and tension (pulling). For instance, with its own weight and the load carried above pushing down upon it, a stone pier is in compression against

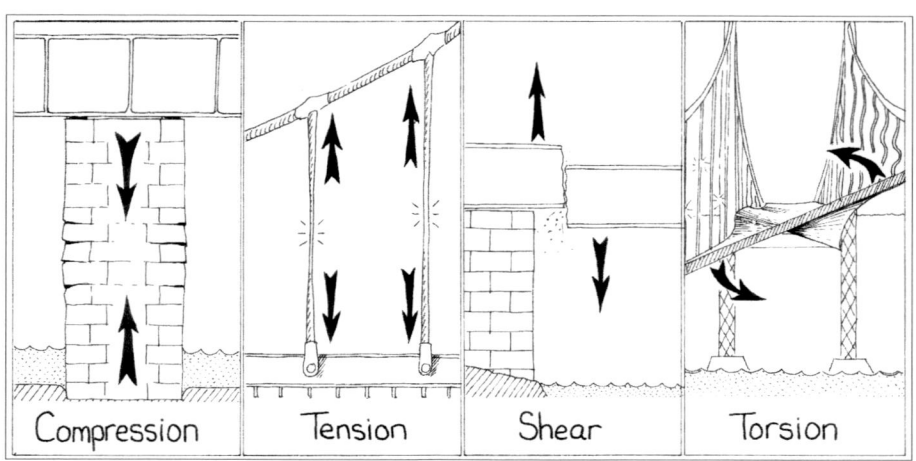

**FIG. 1.7:** *Images showing the four principal forces acting on the material and structure of a bridge, with examples of where they apply (compression acts on the piers; tension on the cables in a suspension bridge; shear on a beam near the abutments; and torsion on the deck of a suspension bridge).*

its foundations. The cables on a suspension bridge, on the other hand, are constantly being pulled down by the weight of the road deck and are hence in tension.

This picture is further complicated for the designer by two other forces which he must allow for. Shear stress results in one part of a solid sliding past another part or shearing in half when under pressure from two sides. If the load on a bridge beam close to the abutment is too great then it might fail in this way. Torsion is the force created by twisting and is best illustrated by the effect wind can have upon the deck of a suspension bridge. Tacomas Narrows suspension bridge in the USA is a famous example, as a photographer was on hand to record the events when a wind of only some 40 knots caused the already notoriously fragile bridge to ripple and twist uncontrollably before disintegrating.

Different materials have different properties in response to these stresses. Hardwoods have both compressive and tensile strength, though not to a great degree. Stone and brick are strong in compression but weak in tension, cast iron is good in compression but brittle in tension, and wrought iron has the opposite properties, but steel (wrought iron with controlled amounts of carbon, chromium, and nickel) is strong in both. High tensile steel, as the name suggests, is designed to withstand even greater stress in tension, but, as it is brittle, it is used in continuous lengths on suspension bridges to avoid joints which might fracture (it is approximately ten times stronger than wrought iron in this situation). Concrete is a versatile material, which has good compressive properties and, when steel rods are added inside it (steel reinforced concrete), it gains the advantage of the metal's tensile strength. Pulling the steel rods with jacks before the concrete set, or inserting wires afterwards, (pre-stressing or post-tensioning), was found to further increase its strength. In addition to its resistance to tension and compression, the material can be formed into a shape or thickened to resist shear and torsion.

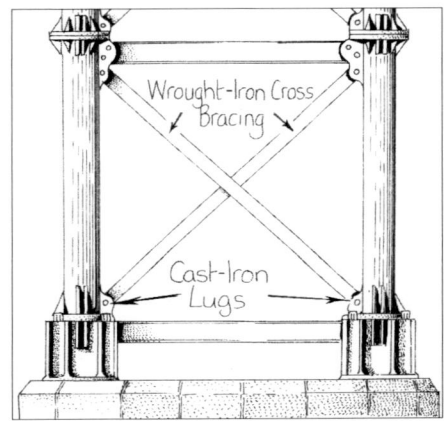

**FIG. 1.8:** *Drawing showing the junction between the cast-iron columns and wrought-iron bracing of the piers on the original Tay Bridge. The cast-iron lugs or flanges to which the bracing was attached were also under tension as the bridge swayed but, as this material is not strong under this force, they should have been strengthened. The failure to do so and the poor quality of the casting were probably the main reason for the collapse.*

# Bridges Explained: Viaducts & Aqueducts

Again the Tay Bridge disaster highlights the problems created when materials are not used in the appropriate manner. The flanges which held the diagonal cross-bracing between the columns were made of cast iron and were not able to take the tensile forces applied to them when stretched. Most of the cross-bracings broke off at these points, and their failure was one of the major reasons for the bridge's inability to withstand the force of the gale.

The periodic leaps forward in bridge building tend to have as much to do with breakthroughs in material technology as in construction techniques or design. The basic types of bridges in this book – arch, beam, cantilever, and suspension – can all be found in ancient civilizations; the only restrictions to their length and load-bearing were the limited materials available. There was only so far a stone arch or timber beam could stretch, and although the appliance of science from the Renaissance onwards pushed the boundaries further, it was the new production techniques of the Industrial Revolution that enabled the next step to be taken.

Iron had been used since before the Romans, but only in small quantities. It was the substitution of coke (coal which has the gases extracted) for charcoal, pioneered by Abraham Darby (1677–1717), that enabled cast iron to be produced on a large scale. Wrought iron was hammered into shape rather than set in moulds, and it did not become available on an industrial scale until the turn of the 19th century. Steel had been used in medieval swords, but it took the development of new smelting processes like the Bessemer's converter (1856) before it could be produced on a scale and to a quality suitable for engineering.

Concrete had been invented by the Romans but the formula was lost over time, and it was not until the invention of artificial cements in the early 19th century that it was rediscovered and began to be used in bridges by later Victorian engineers. It was used at first as a substitute for stone, as large blocks for structural work. It was the introduction of steel rods into its mix which created reinforced concrete and transformed bridge construction from the early 20th century onwards.

The choice of material for a project, however, is more complicated than just choosing the most suitable; costs and logistics must be considered. Stone may be the best material for an arched bridge but it is expensive, as it requires time and skill for it to be hewn from the ground and cut into shape. Transporting it was also time-consuming especially if the quarry was not close to water. Timber was used for most bridges before the 13th century because it was plentiful, easy to transport, and the skills required to build with it were widely available. Stone-arched bridges only become common after this date.

Transport became less of a problem in the industrial age; yet costs still had to be considered. The American railroads used huge viaducts of complex trusses made out of timber (see Fig 1.9) because the thick forests through which they passed provided a

# Building Bridges

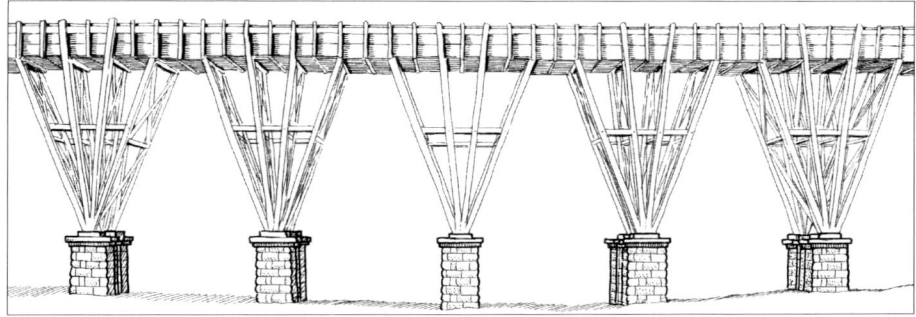

**FIG. 1.9:** *Diagram of one of Brunel's timber viaducts, all of which were replaced in the early 20th century. Brunel used clever arrangements of timber to support the heavy trains and to keep the cost down, as the Baltic timber he used was cheaper than locally available stone. American railroads used similar structures, as readily available wood made their bridges at least a quarter of the cost of their British stone and brick counterparts.*

ready supply that was much cheaper than the more limited metal supplies in the country at the time. Even in this country Isambard Kingdom Brunel used wood rather than stone, brick, or iron when constructing many viaducts on his railway lines in Devon and Cornwall, as Baltic timber was a cheaper alternative.

## Construction

When the engineer had considered the location, the forces the bridge would have to withstand, and the materials which would best resist them, he had to work out how to build his chosen structure. Manpower could be a problem: a stone bridge requires masons, who might be in short supply, and would certainly be relatively costly. A timber structure could be erected by less skilled workmen and would be lighter; so there would be fewer problems with lifting the parts into place, and the foundations could be shallower. Other considerations would include housing for the men and finding suitable areas for workshops, assembling parts, and building machinery.

The first and greatest challenge in the actual construction is to secure a good foundation upon which to build the piers. Across dry land this can be difficult enough, but in the bed of a river or estuary it can be treacherous. It was the Romans who first worked out a reliable way of building in water by using cofferdams. These are enclosures built around the site of the pier base, formed from wooden piles or planks driven into the river bed so that their tops remain above the water level. The water is then pumped or lifted out so that workmen can then dig down through the silt and mud at the bottom

# Bridges Explained: Viaducts & Aqueducts

**FIG. 1.10: MENAI STRAITS SUSPENSION BRIDGE, BANGOR:** *The area in the foreground was an important part of the construction of Thomas Telford's masterpiece. Before the bridge was begun the bank was levelled and tracks laid out to the piers in the picture, with buildings and quays built for the loading and storing of construction materials on the land behind.*

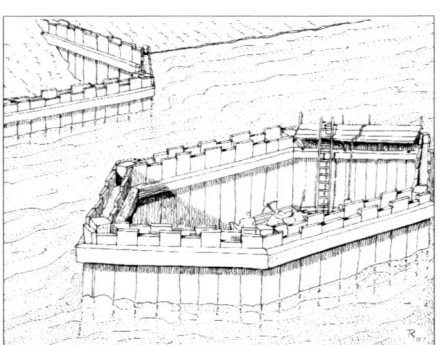

**FIG. 1.11:** *A drawing of a cofferdam showing how the enclosed area can be pumped dry so that work can take place within, before the outer part is removed upon completion. It is not always clear whether the dam was erected on the bank and floated out into position or the individual planks or sheets were pile-driven into position when water levels were low.*

to find a suitable base on which to build the masonry piers. Cofferdams have been used throughout history for building structures in water and are still used today for some projects.

If bedrock could not be found, then piling was used to strengthen the best surface they could find. Tree trunks were driven vertically into the ground to make a close formation (but not touching). It may come as a surprise but most bridges before the industrial age, and many since, had a gridwork or longitudinal arrangements of timber placed on the piles, with the masonry piers built on top. A good hardwood should last for centuries as long as it remains under the water level.

There are limits, however, to the depth to which cofferdams can go and to the strength of water flow they can hold back. As bridges were planned

# Building Bridges

across ever larger and deeper spans of water a new solution was required and the caisson was developed. This is a timber or metal vessel which can be sunk to the floor of the river as masonry is loaded on above. It is a permanent part of the finished structure. There is no standard design, each being developed by the designer to suit the structure and geology of the particular site. The earliest use in this country was for the construction of

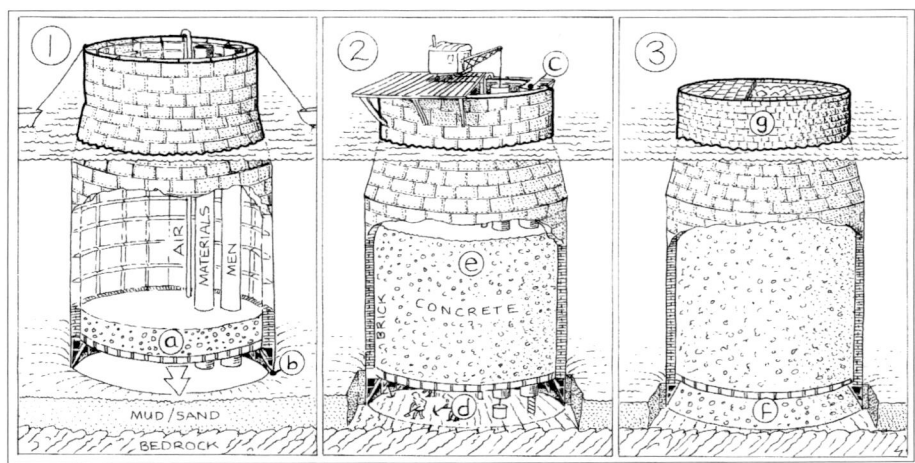

**FIG. 1.12:** *The caisson above is from the construction of the Forth Rail Bridge and is a good example of how the principle worked. The great cylinders were prefabricated and transported to site with the inner framework and false bottom in place. They were then floated out and sunk into position by pumping concrete into the chamber (a) until it reached the seabed. The sharp outer edge of the caisson cut into the sand and mud (b) so that a seal could be made sufficient for pressurized air to be pumped into the lower chamber to clear it ready for men to be lowered down via air locks (c). They would then dig down until a good footing was found, waste being removed by means of a large bucket that could be winched to the surface. Above them an outer skin of brick was built and more concrete poured into the inner core to the top of the permanent section of the caisson (e). The chamber at the bottom was then filled with concrete (f). Inside the temporary part of the caisson at the top an outer wall of granite was built, filled in with stone rubble, and capped off (g). The upper metal section was then removed while the lower section remained, leaving a solid cylindrical pier, one of four which supported each tower. The men who worked in these subterranean chambers were a special breed, as they had to suffer the then little understood bends, in addition to the extreme heat and hard work. Many died in the process, although at the Forth Rail Bridge limited working periods restricted the death toll to two (of possibly 60 to 80 men who died on the project).*

# Bridges Explained: Viaducts & Aqueducts

Westminster Bridge, London, in 1738, when the designer, Charles Labelye, built tall, timber boat-type structures which were floated out to site. The masonry was built up inside, sinking the caisson onto the previously cleared river bed.

In the Victorian period some major bridges were founded with large permanent metal caissons, most in the form of long open-ended cylinders which were sunk to the floor, their lower edge cutting into the river bed. This established a watertight seal, so that air could be pumped into the base to clear it of water, and workmen standing in a gap at the bottom, often no more than 7 or 8 ft high, could dig down to find a good footing while the build up of masonry or concrete above their heads forced the structure further down. Pneumatic caissons were hazardous, the men having to work in a tight space which was unbearably hot, and many died from the bends which, at that time, was little known about, or from drowning when the sealing of the caissons proved faulty so that they filled with water.

Pile-driving machines, cranes to lift stone blocks, and pulleys and scaffolding had to be sourced or designed and built for each specific project. Although Victorian engineers had the benefit of steam-powered machines the medieval mason nevertheless had devices available to him to found the piers and lift heavy items into position. Centring – timber trusses similar to those you find in your loft holding the roof up but with a curved upper surface – were needed to support the underside of the arch as it was being built and had to be purpose-made for each bridge. In the industrial age, however, many bridges were prefabricated away from the site and engineers had to devise ways of shipping them and raising them into place.

All these considerations and technical problems faced by the builder of even the simplest bridge may make us pause and ask why we build bridges in the first place? The cost of building the vast Tay Rail Bridge was huge both in financial and human terms: it took a force of 600 workmen over a period of six years; and in addition to the 75 lives lost on the day of its collapse, a further 20 had died during its construction.

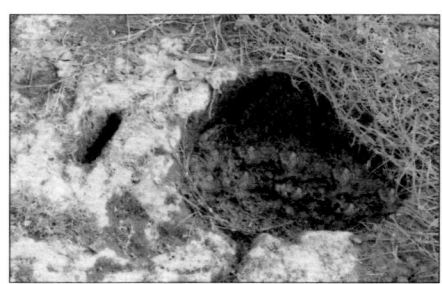

**FIG. 1.13: CHESTERS BRIDGE, NORTHUMBERLAND:** *This stone from the abutments of the Roman bridge which carried Hadrian's Wall across the Tees has a circular hole (right), in which the pivot of a Roman crane used to move the stone blocks was set. Each block had a rectangular lewis hole (left) cut in its centre, into which an iron fitting was placed. This fitting expanded when it was pulled to hold the block fast while it was craned into position.*

Thomas Bouch himself had designed a roll-on-roll-off train ferry some 30 years earlier to avoid building such a bridge, which at the time was seen as impossible. However, as history will show, the speed at which we can travel round the country does matter and the faster route is generally the more popular and can bring prosperity to areas it passes through. The Tay Bridge knocked only half an hour or so off journey times, but in its short life it had already proven its financial worth and this is why it was rebuilt so quickly on an even larger scale.

## Types of Bridges

The only remaining consideration for the designer was which type of bridge to build. Different structures have been popular at different times, depending on the knowledge and materials available. The earliest types probably developed either from stepping stones over which were placed large slabs to make clapper bridges or from fallen logs which were secured to make permanent timber structures. Despite popular images of barbarians, British pre-Roman society was actually quite sophisticated, trading with the continent, and creating beautiful works of art and even bridges. Simple timber structures consisting of logs driven into the river bed with planks laid across enabled smaller rivers to be crossed. Beam bridges and later examples are covered in Chapter 6.

This form of bridge was used on a much wider scale by the Romans after their invasion in AD 43. The largest examples were built by the army as part of their massive road network and used stone piers with wider decks to carry carts and legions. It was the departure of these soldiers in AD 410 and the subsequent collapse of the market system which greatly reduced the need for such bridges in the so-called Dark Ages. The Romans, however, had left one form of structure which was new to

**FIG. 1.14:** *Stone arch bridges from the Middle Ages to the Georgian period could range from large multi-arched structures for main road crossings in busy towns and cities, like this example in Hereford (left), to narrow single-arched bridges used for horses and mules carrying packs (hence they only needed a narrow passage, often with no parapet), as in this example at Three Shires Head near Leek, Staffordshire (right).*

these shores: the arch. Although only a few arched bridges from the Roman period are known to us, the form was revived in the 12th and 13th centuries as trade and wealth increased and it remained dominant until the Victorian period (see Chapters 2 to 5).

The stone structures were popular and beam bridges were still being built, but it is likely that until the 18th century most rivers and streams continued to be crossed by fords and ferries. At that time, with the increased demand for long-distance travel and the arrival of canals and then railways, bridges became widespread and new forms were experimented with. As iron became available on an industrial scale, new girder and trussed bridges were developed, capable of supporting the heavy load of a canal or a moving train (see Chapter 6). The same advances produced the answer to building bridges over large expanses of water without obstruction to shipping: the suspension bridge (see Chapter 8). Smaller watercourses with the same problem could be crossed with the aid of new forms of moving bridges which either swung or lifted the deck out of the way (see Chapter 9). With the advent of steel, another ancient form of bridge was revived: the cantilever bridge. Examples of these and modern concrete types are described in Chapter 7.

It was not only waterways, rails, and roads which needed to be crossed, the need for canals to be kept at a level across valleys to avoid building expensive locks down either side resulted in the use of aqueducts. Some of the most breathtaking examples are explained in Chapter 11. A similar problem faced railway and modern motorway engineers and the viaducts they built are covered in Chapter 10.

FIG. 1.15: PONTCYSYLLTE AQUEDUCT, TREFOR, WALES: *There are few more terrifying structures to walk across than the Pontcysyllte aqueduct. Pedestrians have an iron rail to protect them from the 120 ft fall, but boats have nothing between themselves and the raging River Dee below!*

CHAPTER 2

# Roman and Medieval Arched Bridges

**FIG. 2.1: LONDON BRIDGE:** *An impression of how the first stone London Bridge may have appeared in the 16th century. It had been designed by a priest, Peter of Colechurch, who died shortly before its completion in 1209 and was buried in the crypt of the chapel which stood upon it. His structure survived for more than 600 years, but when it was demolished in the 1830s his bones were uncovered and are believed to have been discarded into the Thames by workmen.*

'London Bridge is falling down, falling down, falling down.' The well-known nursery rhyme of which this is the opening line is almost certainly of ancient origin and probably refers to an earlier bridge than the first stone structure in the picture above. The Romans had built a timber bridge across the Thames in this area, and this or a later one was damaged by fire and was replaced by a similar structure in 1169. It is not clear why, but less than ten years later the new stone bridge was begun. It was 900 ft long, with huge stone piers sunk into the tidal river. These were almost as wide as the arches or spans backing the river upstream of it, and had the effect of slowing the flow enough for the water to freeze over in winter and at the same time creating such a strong current below that the river bed was scoured out to a depth ideal for shipping in what it still known as the Pool of London.

What must have appeared most striking at the time was not just the scale of the project – it took over 30 years to build – but the imposing row of nineteen stone arches. Previous structures had used timber beams, and stone-arched bridges were rare at the time and unknown here on this scale, yet the bridge's superior load-bearing capacity meant it could support the community of houses and shops which grew up on it and came to characterize it.

London Bridge had been rebuilt to maintain the capital's position as a trading centre, and, although many medieval structures were constructed with convenience in mind, most were usually concerned with keeping traffic moving to market. Towns could grow or fail depending on whether they had a permanent crossing point.

After the Romans landed in AD 43 the first roads and major bridges were built by the army to ensure that legions could move swiftly to any corner of the empire's new acquisition in order to quell uprisings. Their main use, once order had been established, was trade – the movement of grain, goods, and materials to local markets and foreign ports – which had been one of the main reasons for the invasion in the first place. It was the loss of this national market network when the last legions left these shores in AD 410 rather than the influx of Germanic tribes which in part distinguished the supposed Dark Ages from the preceding times. Most of the Roman roads and bridges fell into disrepair because the country fragmented into smaller tribal groups, so that long-distance trading routes were used less. In some cases new centres were established away from the existing roads. Some bridges were maintained, and new ones were constructed but these were timber structures with horizontal beams supported on vertical piles or stone piers (see Chapter 6).

It was not until a more unified kingdom emerged in the 10th century, and then an increase in trading and the founding of new markets in the 12th and 13th centuries, that new more permanent bridges were deemed necessary. Some were built to replace earlier timber structures; others were constructed as part of a new road or town, with the intention of diverting business away from a neighbouring market. Most goods were transported by horse or mule, and so medieval bridges were generally only wide enough to let a single vehicle pass. (Nearly all, though, have subsequently been widened.)

# Roman and Medieval Arched Bridges

**FIG. 2.2: ST IVES BRIDGE, CAMBS:** *This early 15th-century bridge with its original chapel (see Fig. 2.21) replaced an earlier structure which was built around 1100. It played an important part in making what was previously a small village and priory into an important medieval town and was possibly constructed and financed by the abbot of the mother house.*

There was no controlling transport body in the medieval period and bridges were built by a variety of individuals or groups. Many were paid for by the church, monastic orders, town or city guilds, burgesses, or wealthy lords, while some were funded collectively by the local nobility who would benefit from it. The Crown, however, was rarely involved in bridge building. Deciding who was responsible for maintenance once it was completed was another matter. If the original builder could not be established, then responsibility fell upon the local parish and, if the money could not be raised from wealthy individuals, groups, or charities, then a grant of pontage would enable tolls to be charged for any necessary work. Many had chapels upon or near the structure, where mass could be chanted for the deceased, or indulgences granted by the church for remission of sin, in return for financial contributions towards the bridge's upkeep. In other cases rent from land and buildings was bequeathed for this purpose.

The stone-arched bridges of this period varied in size from small structures for packhorse traffic to large, multi-arched examples across major rivers. However, there were only two main types of arch used: semi-circular and pointed.

## SEMI-CIRCULAR OR ROMAN ARCHES

Although possibly first devised by the Sumerians as long ago as 4000 BC, it was the Romans who first developed and widely used a semi-circular arch, although in bridges in this country it was rare. It is in medieval structures that the earliest examples will be found, comprising wedge-shaped stones

# Bridges Explained: Viaducts & Aqueducts

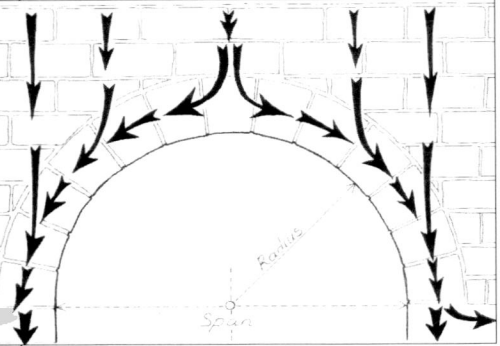 

**FIG. 2.3:** *A diagram showing a semi-circular arch. The arrows show how the weight of the structure above and the load it carries are transferred out towards the piers and sides of the bridge.*

**FIG. 2.4:** *An example of a semi-circular arch with a single ring of voussoirs; the chamfered edge on the lower side was a feature of later medieval bridges.*

(voussoirs) arranged in a half-circle with the narrow ends on the underside. (Alternatively, regular shaped stones could be used with mortar filling the gaps.) The downward pressure from their own weight and that of any load built on top forces the stones tightly together, in effect transferring the force from above down the sides and onto the point from which it springs. A small force is transferred to the sides, however, and so the designer also has to bear this in mind (see Fig 2.3). This simple and ingenious design enabled the Romans to bridge large gaps with small sized blocks and to create more lasting and stronger structures. The arch is so durable that it can survive movement in its foundations or abutments to such a degree that one, two, or even three cracks or hinges in its line will not lead to collapse; only when a fourth appears will it fail.

A problem with a semi-circular arch, however, is that it does not achieve its strength until the central voussoir (the keystone) is put in place. This means that the arch needs support from beneath while it is being constructed. A timber frame in the shape of the finished arch, called centring, was used as temporary support, to be removed carefully when the arch had set.

## Pointed or Gothic Arches

The disadvantage of the semi-circular arch is the inflexibility of its height-width ratio. The height of the arch is around half the width of the span (see Fig 2.3); so, if you are building a bridge and need to make the gap between piers wider, then the height has to increase too. This is a problem if poor conditions in the river bed mean that the piers have to be founded at irregular intervals, resulting in a bridge with high

# Roman and Medieval Arched Bridges

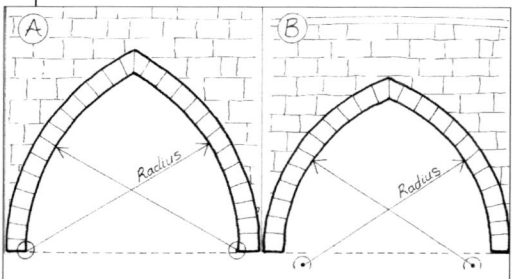

**FIG. 2.5:** *A diagram of two pointed arches, showing how by moving the centre points of the radius the opening can be made wider or taller.*

**FIG. 2.6:** *A pointed arch with a double chamfer edge.*

and low arches undulating like a wave. Alternatively, if greater clearance is required under one of the spans, the width has to be increased in proportion, something which builders in the medieval period rarely did. It was not until the late 12th century that masons began to experiment with stone-arched structures that broke away from the Roman model.

The main development was the pointed arch, which appeared here in monastic churches from around 1170, although the form had been used elsewhere even in pre-Roman times. The advantage of the two-centred arch is that it can be varied in width without having to increase its height, and likewise it can be raised to give greater clearance without altering the span. This flexibility made it a popular form for bridges in the 14th and 15th centuries.

Alongside the new form was the introduction of rib vaulting. Rather than the entire arch being load-bearing, a series of beams or ribs in stone supported the whole so that the filling behind could be reduced in thickness. This was useful to bridge-builders, as it reduced the overall weight of the span, thereby reducing pressure on the piers. As only the ribs needed support during construction, the cost and time of building was reduced. Ribs can be found on both types of arched bridge. There were nevertheless limitations on the width of the span of this type of arch, and both semi-circular and pointed arch medieval bridges have numerous piers to cross even the most modest river.

## BRIDGE DESIGN AND CONSTRUCTION

### Piers

The piers carried the force transferred down through the arch and to a lesser extent from the sides of semi-circular

FIG. 2.7: *The stone ribs with chamfered edges on the underside (soffit) of a medieval bridge. Note that the stonework between the ribs is of smaller, poorer quality stonework, a cost-saving measure which is possible as the ribs take most of the load.*

and pointed arches. Although medieval masons could evaluate some of the basic requirements to support them, bridge builders tended to play safe and build large piers except occasionally where bedrock was prominent. An advantage of large piers and short spans was that when constructing a long multi-arched bridge, a single arch could be built and then left until conditions, funds, or materials were in place to build the next one, an important consideration in a structure like London Bridge, which took around 30 years to build.

The downside of bridges with conservatively short spans and numerous piers was that they obstructed the flow of the water. This in effect dammed the river on the upstream side, increasing the chances of flooding and damage to the structure. The rush of water between the piers caused a hazard to river traffic and intensified the scouring, which undermined the piers. To protect the base of the piers, piles were driven round them and the gap filled in with rubble to create a boat-shaped guard, called a starling.

As there has been limited study of the methods used in bridge construction in the medieval period compared with other building types, an element of guesswork is involved, especially in respect of the foundations. There would have been regional variations and different methods, depending upon the mason in charge and the geology of the site. In most examples some form of dam must have been established around the base of the piers to keep the water and mud out as the site was excavated. Work may have been limited to times of low water levels such as summer or when the tide was out. On major structures, especially in midstream, more permanent timber-piled cofferdams may have been used. Some were built directly upon bedrock, others had generally shallow foundations with timber piles driven into a compacted mud base to support the masonry. The pier was built up on this, usually with just a single step at the bottom. (Later arched road bridges usually have more in order to spread the load.)

The remains of piers, abutments, and parapets from the few Roman stone bridges so far found have holes in which would have sat an iron clamp, held in place by lead, an unusual

# Roman and Medieval Arched Bridges

method of fixing them together (although these are also found holding coping stones along the top of later stone parapets). Medieval bridges were built with mortar, which was a mix of lime, sand and water. Piers were generally built as a hollow shell infilled with rubble and cement.

**FIG. 2.8:** *This cut-away view of a large stone multi-arched bridge shows how it may have been constructed in the late medieval period. Firstly the foundations had to be established behind a cofferdam or a more basic wooden barrage or sand bags (1). A base of vertical timber piles (the staddle) was driven into the bed when building on soft ground as a base for the masonry pier. An outer ring of piles, called the starling, was also used to protect it from scouring. The piers were usually hollow and filled with any available rubble and mortar (2), and then the arches were built over wooden centring (3). The spandrel walls would then have been constructed, and the roadway formed above an infill of rubble perhaps with a clay lining to keep water away from the arch (4). Wooden railings or none at all were likely on early bridges, stone parapets being more common later.*

# Bridges Explained: Viaducts & Aqueducts

FIG. 2.9: *Medieval pier bases were usually plain with just a single step, as in this example from Stone, Staffs, which also has a recent concrete skirt built around it for protection.*

FIG. 2.10: *Much of the masonry on the few Roman arched bridges found in Britain was held in place by iron clamps set in holes filled with lead. A similar method was later used to fix coping stones along the top of parapets.*

FIG. 2.11: *The remains of sockets (left) and corbels (right) which supported the wooden centring used during construction of the arches.*

## Arches

The centring for the arch was built up by carpenters on stone brackets (corbels) or sockets in the walls of the piers, just below the point from which the arch was to spring. The corbels and sockets can sometimes be found in medieval bridges.

The arch voussoirs could be built without mortar between them and nevertheless maintain a strong bond, but, because even the smoothest stone has a finely rippled finish, only the tips of the undulations are in direct contact with the neighbouring pieces. Mortar is used in walls and arches, not just for adhesion, but to fill the gaps in the surfaces so that the downward force being transferred through each block is passed on evenly over the surface. The mortar also reduces the chance of the blocks slipping horizontally over each other. Because of the nature of the forces in an arch with rib vaulting, it had the added advantage of not requiring so much fine masonry, only the ribs needing to be cut accurately, which thus reduced the building cost.

# Roman and Medieval Arched Bridges

**FIG. 2.12:** *The voussoirs on the left-hand example are cut to shape and of good quality despite erosion on some. The right-hand example is of poorer quality, having rectangular pieces of different sizes with mortar and small pieces of stone filling in the gaps.*

The space behind the spandrels and on top of the arches which supported the road was usually made up of rubble and other waste material to form a solid base for the finished road surface. It was important to stop water penetrating into the joints in the arches and so a lining of clay may have been used to make it waterproof.

## Cutwaters

Early examples, especially small structures, may have had a flat front facing the river flow, but most medieval bridges that stand today have a triangular or rounded cutwater to guide the water through the spans. These were usually extended upwards to create a pedestrian refuge along the narrow deck. Cutwaters on both sides of the bridge tend to appear on later examples or could have been added at a later date to an earlier structure. Many of these refuges have since gone, as it was convenient to build over them when widening the bridge at a later date.

## Parapets

Despite seeming dangerous to modern eyes, it is likely that many medieval bridges had no parapet to prevent travellers being knocked off the bridge (which was recorded as happening on a number of occasions). Some may have had timber railings cantilevered off the sides, but most stone parapets are later additions. Drainage was also an important consideration: water on the surface was a hazard to travel and could get into the masonry and do damage in

**FIG. 2.13: HUNTINGDON BRIDGE, CAMBS:** *The triangular cutwaters between the arches on this bridge dating from the late 14th century are typical of large medieval bridges. They are large and plain and extend to the full height of the bridge, creating a refuge from traffic for pedestrians on the originally narrow carriageway.*

# Bridges Explained: Viaducts & Aqueducts

FIG. 2.14: TILFORD BRIDGE, SURREY: *One of a number of 13th-century bridges in the Wey Valley which are believed to have been built by the monks from Waverley Abbey. The timber parapets are modern but are probably similar to medieval types. The rounded cutwaters are also rare, although on the upstream side they are the more usual pointed style.*

winter. Drain holes, often with an extended chute so that the water did not pour straight down the front of the bridge, were usually fitted through the base of the parapet walls (see Fig 2.16).

## Ornamentation and Decoration

Despite the glorious decoration lavished upon the cathedrals and abbeys of the

FIG. 2.15 (*above*): MEDBOURNE BRIDGE, LEICS: *This packhorse bridge, believed to date from the 13th century, has modern timber parapets which give an impression of how many medieval structures may have originally appeared. Like many packhorse bridges, it has an adjacent ford with goods being carried over the bridge when the river was too strong.*

FIG. 2.16 (*left*): *Close-up of a drain outlet on the parapet, through which rainwater was directed off the deck to protect the infill and arch below.*

# Roman and Medieval Arched Bridges

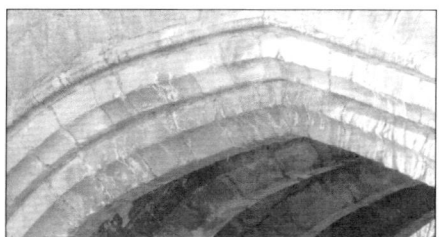

FIG. 2.17: *Decoration was rare on medieval bridges but does appear in some later examples, as in these two 14th-century bridges at Thornborough, Bucks. (top) and Huntingdon, Cambs. (below).*

FIG. 2.18: TRINITY BRIDGE, CROWLAND, LINCS: *This unusual three-legged bridge originally spanned the junction of rivers which ran down the centre of the streets. It was probably built for the nearby abbey and has decorated mouldings around the arch rings. Its most notable feature today is the seated statue on the far right, which is believed to have been moved here from the ruined abbey in the early 18th century.*

age, the medieval bridge, hewn by the same hands, tends to be purely functional. Buildings on bridges, especially chapels, were suitably ornate, but there seems to have been little time spent on beautification of the structure. One exception was the edge of the arch, which in early examples tends to be sharp but, in later examples, was chamfered, often with two or even three bands.

Originally there would have been a cross on most bridges, although none survive, having been demolished by protestant zealots, with but a few bases left as possible evidence. An exceptional bridge with three arms (built over a junction of rivers which used to run down the centre of the streets) still stands in Crowland and features a statue, although it is likely that this was resited here at some later period.

## BUILDINGS ON BRIDGES

Another feature of Roman and medieval bridges of which little evidence remains today is that of buildings upon or next to the structure.

## Chapels

Religious associations with bridges date back into prehistory. Water has held a special place in man's eyes, and valuable metal artefacts dating from the Bronze and Iron ages have been found placed in rivers alongside contemporary timber piled structures. Altars dedicated to Neptune and Oceanus were found at the site of a Roman bridge, indicating that there may have been a shrine upon it.

Reverence for structures crossing rivers continued into the medieval period, when chapels were built on or at the ends of many of the largest bridges. Not only did they reflect the sacred crossing but also the fact that many bridges were paid for by the church and they were points where tolls or contributions towards the upkeep of the structure could be collected in return for a blessing. Some were chantry chapels which were paid for by local dignitaries in return for the daily singing of mass for the salvation of their souls after death.

FIG. 2.20: ROTHERHAM CHANTRY BRIDGE: *Founded by the Archbishop of York, the chapel and bridge were built in 1483. Since then the chapel has been used as a prison and shop before it was restored in 1924. The window tracery still shows marks made by Cromwell's soldiers when hammering out the glass.*

FIG. 2.19: WAKEFIELD CHANTRY BRIDGE: *This chantry chapel was built in the mid-14th century and endowed by Edward III in 1358. It was heavily restored in 1847 and 1937, and is still used for services.*

Bridge chapels stopped being used as such after the Reformation in the 16th century, and many were damaged as a result of puritanical fervour during and after the Civil War. Those which remained ended up as prison cells, shops, pubs, and private houses, although a few were again used to hold services in the last century. They survive today in St Ives (Cambridgeshire),

# Roman and Medieval Arched Bridges

**FIG. 2.21: ST IVES BRIDGE, CAMBS:** *The interior of the chapel (top) and the basement (below) of the structure featured in Fig. 2.2. The building was converted into a house in the 18th century and was only returned to its original form in 1930. Although a small space for worship, it was at least under one roof. The chapel at Droitwich (long since gone) was built in two parts with the priest on one side and congregation on the other and the road running through the middle!*

Rotherham, Wakefield, and Bradford-on-Avon, and in part at Derby, Cromford, and Rochester.

## Gatehouses

Other medieval bridges served as part of a town's or castle's defensive walls, or, especially near the borders of Wales and Scotland, were in a position where a gateway was deemed necessary. Like chapels, the towers could be built on a pier or island along the bridge or at one

**FIG. 2.22: MONNOW BRIDGE, MONMOUTH:** *The top view shows this famous bridge today; the tiled roof on the gatehouse dates from the 18th century, when it was converted into a house. The lower drawing shows how the bridge and gatehouse may have originally appeared in the early 13th century.*

# Bridges Explained: Viaducts & Aqueducts

inhabitants. Another notable example was in Bristol, although both are long since gone. Only at Lincoln, where later 16th-century houses still stand on the partially-rebuilt single span bridge, can you get a good impression of how they would have looked. Some bridges, however, still have buildings encroaching over the approaches or first arches (see Fig 2.33 Durham).

FIG. 2.23: HOLT BRIDGE, CHESHIRE: *There was originally a gatehouse on this bridge, which is on the border between England and Wales, and the bottom corner of a window from it still remains in the parapet; it can be seen in the top right of the photo.*

## Mills
Mills were often associated with medieval bridges, the numerous piers

end of it, giving the added advantage of being a place to collect tolls for passage. Only two survive today, at Monmouth and Warkworth, but evidence of where they once stood can occasionally be found on other bridges.

## Houses and Shops
Houses and shops were also a distinctive addition to some of the largest bridges, the rents being used in part to pay for repairs to the bridge. They were usually cantilevered off the structure so that they partly hung over the edge on wooden beams or were supported on walls or posts resting on the starlings or islands in the river. One of the most famous examples, seen in old prints, is the rambling collection of timber-framed buildings along the medieval London Bridge, which became a self-contained community in the eyes of its

FIG. 2.24: LINCOLN HIGH BRIDGE: *The only medieval bridge in the country with buildings remaining on it. The current structure, on the much widened and altered arch over the River Witham, was built in 1540, with a stone lower storey and timber-framed upper storeys.*

raising the water level to provide a head of water which could be channelled to power a waterwheel. Surviving examples are found on banks above or below bridges, although a few were built on the bridge itself.

FIG. 2.25: DUDDINGTON BRIDGE, NORTHANTS: *The turbulent water in the foreground comes from the old mill built just upstream of this bridge.*

## EXAMPLES OF STONE-ARCHED BRIDGES

## Roman Bridges

FIG. 2.26 *(left)*: CHESTERS BRIDGE, HADRIAN'S WALL, NORTHUMBERLAND: *The top photo shows the remains of this Roman stone-arched bridge dating from the 3rd century. The lower drawing illustrates how it may have originally appeared; the surviving stone work of the abutment and lower part of the gatehouse is highlighted in thicker lines.*

FIG. 2.27 *(below)*: WILLOWFORD BRIDGE, HADRIAN'S WALL, NORTHUMBERLAND: *The remains of sluices in the eastern abutment may have been for a mill or served as flood relief channels.*

## BRIDGES EXPLAINED: VIADUCTS & AQUEDUCTS

## Medieval Road Bridges

**FIG. 2.28: LLANGOLLEN ANCIENT BRIDGE, LLANGOLLEN, DENBIGHSHIRE:** *Although restored and widened on numerous occasions, this bridge built in 1282 still retains much of its medieval form.*

**FIG. 2.30: THORNBOROUGH BRIDGE, BUCKS:** *This 14th-century structure still retains its typically bulky medieval form with full height cutwaters on the upstream side only (see close-up of arch rings in Fig. 2.17 and also Fig. 3.10).*

**FIG. 2.29: HARROLD BRIDGE, BEDS:** *The hotchpotch nature of this bridge is due to different local bodies or lords having been responsible for individual arches! Also of note is the long causeway, which originally carried pedestrians over the flood plain approaching the structure. Peer beneath and you can see the progressive widening of its arches.*

**FIG. 2.31: HOLT, CHESHIRE:** *Built around 1338, this sandstone structure stands on the border between England and Wales and originally had a gatehouse, where tolls had to be paid for crossing (see Fig. 2.23). It was rare for these to see action, but this one was called into service in 1643 when Roundheads attacked Cavaliers, trying to prevent them cutting off Chester, although they eventually broke through after laying a decoy.*

# ROMAN AND MEDIEVAL ARCHED BRIDGES

FIG. 2.32: BAKEWELL BRIDGE, DERBYS: *A 15th-century bridge which despite widening retains its original character. It is likely that a regular set of arches like this means it was built and maintained by a single body rather than by many (compare Fig. 2.29).*

FIG. 2.34: NUN'S BRIDGE, HINCHINGBROOKE, HUNTINGDON: *This medieval bridge, which according to local legend is haunted by a nun, possibly owes its differing arches to having been split between two parishes, each responsible for one half.*

FIG. 2.33: OLD ELVET BRIDGE, DURHAM: *One of the earliest surviving arched bridges, it was begun in around 1160. It had buildings across its entire length in the medieval period, including two chapels. Some later houses still stand on the land arches as seen in the right of the picture.*

FIG. 2.35: OLD DEE BRIDGE, CHESTER: *This large sandstone structure dates from the late 14th century, apart from the later wide arch on the right, which was originally the site of a tower and drawbridge. An early example of a water-powered electricity station stands on the site of the ancient mill on the far left of the photo.*

# Bridges Explained: Viaducts & Aqueducts

**FIG. 2.36: BROMHAM BRIDGE, BEDS:** *This is a medieval stone-arched bridge across the River Ouse of unknown date. Like the one at nearby Harrold, it too has a long stone causeway, originally to carry pedestrians, while carts had to use a track and then rise up a ramp onto the furthest right arch in the photo. This didn't help a poor woman who fell in during the winter of 1281 and was carried on an ice flow as far as Bedford, whereafter she vanished.*

**FIG. 2.38: BIDEFORD LONG BRIDGE, DEVON:** *This massive 680 ft bridge over the tidal River Torridge was built in around 1315 with 24 pointed arches of varying width, presumably so that the piers could be sited on the best footing.*

**FIG. 2.37: STURMINSTER NEWTON, DORSET:** *A late medieval bridge over the River Stour, it was widened by extending the arches between the cutwaters, leaving only their outer point protruding.*

**FIG. 2.39: CULHAM BRIDGE, OXON:** *An early 15th-century bridge, it was built as part of a medieval road improvement scheme undertaken by a local guild, which also included a causeway and a larger bridge over the Thames at Abingdon. The result was that traffic and business was drawn to this town and away from other crossing points.*

## Packhorse Bridges

FIG. 2.40: WYCOLLER, LANCS: *This unique packhorse bridge was probably distorted during construction in the late medieval period, as the arches, made of voussoirs running the full width, were sprung directly from the rock without the bank being levelled first.*

FIG. 2.42: WEST RASEN, LINCS: *This tiny sandstone bridge dating from the late medieval period is in virtually its original form, with parapets only two feet high (although these are often later additions), and no later widening.*

FIG. 2.41: SUTTON BRIDGE, BEDS: *This simple, fine quality structure is of unknown date. It was found to have been built upon four long elm beams running under both arches, and radio carbon dating has suggested the bridge was erected in the late 13th century. As with most packhorse bridges, the ford in the foreground was used for most traffic except when the river was too high, to save wear on the bridge.*

## CHAPTER 3

# *Tudor and Stuart Arched Bridges*

**FIG. 3.1: SONNING BRIDGE, BERKS:** *This red-brick structure, with its simple semi-circular arches and humpback form to enable boats to pass beneath, is little different from the bridges covered in the previous chapter. Yet this was built in 1773. It shows how the development of bridges in this country had in many examples hardly moved on from the medieval period.*

Despite the glory of our medieval cathedrals and castles, which are comparable with much on the other side of the Channel, our bridges could be clumsy, bulky, and technologically limited compared with the best in Europe. Just as the longest of our stone-arched structures, London Bridge, was nearing completion at the end of the 12th century, a far more daring and elegant bridge had already been finished across the Rhone at

# Tudor and Stuart Arched Bridges

Avignon, with spans of 100 feet, three times those of London. In 14th-century Florence the problem of piers obstructing river flow and of inconvenient hump-backed roadways was solved on the Ponte Vecchio with the daring use of shallow segmental arches with 100 ft spans and only two piers, yet wide enough to support houses and shops.

As the Renaissance blossomed on the Continent a new breed of designer emerged who used mathematics and engineering theory rather than experience and experiment alone to stretch the limits of stone arch construction. The rebirth of classical art and architecture had little effect on these shores; we became rather insular and backwards in many respects, cut off after the Reformation from the main flow of continental style. This is clearly evident in the bridges built here in the 16th and 17th centuries. There were occasional glimpses of what could be done, but usually only when foundations and abutments were secured on rock. Otherwise the design and construction of bridges was not far beyond that of London Bridge five hundred years earlier.

Part of the problem was the lack of co-ordination and large-scale financing of the road network. The first attempts to improve their state came with the Highways Act of 1555, which reinforced the existing practice of parish responsibility for local road maintenance in the wake of the Dissolution of the Monasteries. Little changed though, as improvements were often counteracted by increased use, as more goods were transported by waggon, cart or packhorse and carrier services between towns. We were now a more insular nation, looking to protect our borders rather than gain foreign acquisitions; and royalty, gentry, and merchants spent more time travelling between towns and country seats than before, aided by new road maps and improved carriages.

Bridges suffered a similar fate, as, with the loss of finance from monastic groups, chantries, and indulgences after the Dissolution of the Monasteries and the Reformation, legislation put the responsibility for their building and maintenance upon the borough, county,

FIG. 3.2: DEVIL'S BRIDGE, KIRKBY LONSDALE, CUMBRIA: *An impressive and imposing structure from the late 15th century, it demonstrates that the knowledge existed to build large spans with narrow piers and could be built where the foundations were solid. It still has the medieval ribs, cutwaters on both sides, and three chamfered arch rings, which are typical of this date.*

# BRIDGES EXPLAINED: VIADUCTS & AQUEDUCTS

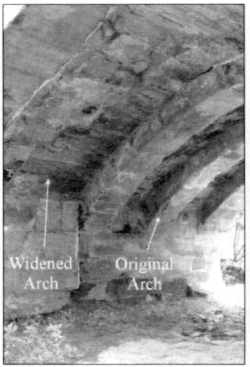 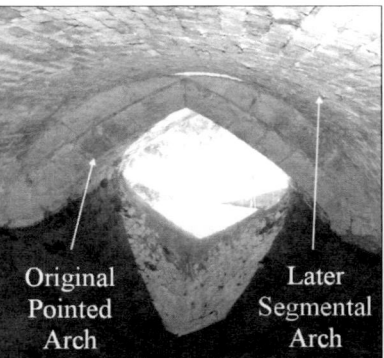 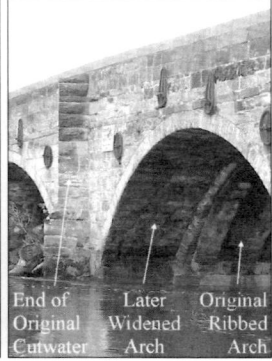

**FIG. 3.3:** *It is in this period that the process of widening bridges to accommodate two lines of traffic begins, a process which increased in the 18th and 19th centuries and continued well into the 20th. Widening can usually be detected by looking under the arches, where there will be a division between the original build and later extensions (left), or sometimes the shape of the arch is also different (centre). The widening was usually done by extending out between the cutwaters so that just their short pointed end remained (right). In some the original face was retained and reconstructed after the structure had been widened; in other situations a completely new form was used (see Fig. 4.2).*

or parish. Despite the increase in traffic, the minimum was often done to keep them open. However, many people did become wealthy on the back of the break-up of the monastic estates and a general rise in trade and income from rents, and it was they who often paid for new structures and maintenance of the old.

## TYPES OF BRIDGE ARCHES FROM THIS PERIOD

The pointed medieval arch and ribbed vaulting fell from favour, although a variation, the four-centred arch was sometimes used in the 15th and 16th centuries. This springs from two small arcs on either side and then two larger

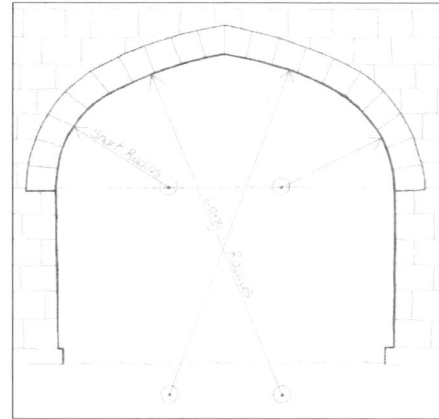

**FIG. 3.4:** *A four-centred arch comprising two short radii on the springing line and two larger ones below it.*

ones on the upper side which meet at a shallow point in the middle. The semi-circular arch, which had always been used to some extent, returned to favour in the 16th and 17th centuries as Classical styling and Roman structural forms began to be appreciated.

## Segmental Arches

The Ponte Vecchio had shown in the mid-14th century how, by using a segment of a large semi-circular arch, a more convenient flat and wide span could be created, reducing the need for numerous obstructive and costly piers. This form of segmental arch began to appear in England in the 15th century and became the principal form of arch used during the Tudor and Stuart periods.

The disadvantage of this form of arch is that an increased force from the dead and live loads is now transferred sideways, pushing the supporting piers and abutments outwards. This should not be a problem with a single span bridge, although the abutments would have to be larger than in earlier forms or fixed on bedrock. However, in multi-arched bridges a completed arch could force over the pier of an uncompleted one. In the more adventurous examples this was solved by building the arches simultaneously so that when the temporary centring was removed the arches would apply their lateral force along the line of the bridge and out into the banks. With most examples from this period, however, it seems that the builders were more cautious and tended to continue the medieval tradition of building substantial piers which could resist the outward force, thus making construction easier but rather defeating the point of using segmental arches. The ratio of the span to the width of the pier, which was generally 1:1 in medieval bridges, did expand to 2:1 and sometimes more, when bedrock was near the surface to give a more robust foundation.

## Design and Construction

There were few major steps forward in the style of structure and its construction until the Georgian period, although even then there were bridges built in a more or less medieval form (see Fig 3.1). The use of starlings around the base of piers died away in this period. The more usual way of tackling scouring was by constructing an apron of stones or cobbles around the base. Piers still tended to be large and numerous across most major rivers,

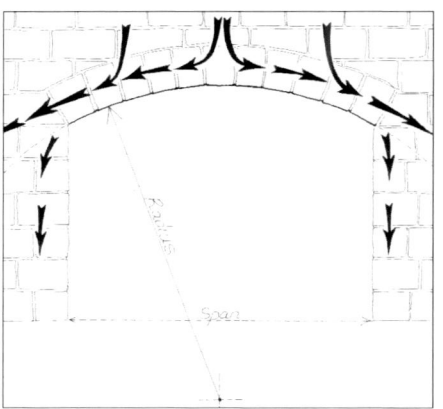

FIG. 3.5: *A segmental arch with arrows showing the direction in which the loads are transferred.*

# BRIDGES EXPLAINED: VIADUCTS & AQUEDUCTS

FIG. 3.6: SHEEPWASH BRIDGE, ASHFORD IN THE WATER, DERBYS: *Although many structures may appear rustic and of medieval origin, as with this packhorse bridge, the use of segmental arches on even the smallest example usually means it dates from the 15th century at the earliest, and more usually from a later period.*

usually with pointed cutwaters on both sides. Often these were cut back to form three-sided (half hexagonal) refuges above, a distinctive feature of this period. Bridges continued to be plain and functional in appearance. The occasional use of a decorative string course or ornamentation was usually found only on a private estate, to complement the owner's new house and landscaped garden.

In this period brick began to be used for bridges for the first time, although

FIG. 3.7: DOVE BRIDGE, UTTOXETER, STAFFS: *Many bridges were simply repaired rather than replaced, and it is common to see a mix of arch types in one bridge. In this example the outer pointed arch (left) is from the original 14th-century structure, but the two main semi-circular ones (right) are replacements dating from the late 17th century; presumably the originals suffered from flood damage. Segmental arches are also found as replacements for ones damaged in floods or the Civil War (e.g. Fig. 2.2, far left) or fill a gap where a drawbridge once crossed (e.g. Fig. 2.35, far right).*

initially it was usually used only for patching up or extending an existing structure. Although introduced by the Romans, the method of making bricks had been lost on their departure. It was reintroduced here from the Low Countries in the 14th century. The hand-made bricks were baked on site and were a luxury product for the finest of houses. By the late 16th century they began to be produced on a larger scale and to more standard sizes, and local brickworks began to be established, but it would not be until the late 18th century that their use would become more widespread.

### EXAMPLES OF STONE-ARCHED BRIDGES FROM THIS PERIOD

## Packhorse Bridges

**FIG. 3.8: HEBDEN BRIDGE, YORKS:** *With growing trade in the 16th and 17th centuries, but a still limited transport system, there was increased use of packhorses to carry goods across country, especially in upland areas where navigable rivers could not reach. It is in this period that many of the finest stone-arched packhorse bridges were built. Except for the heightening of the parapet in 1890, this example has changed little in form from when it was originally built in 1510.*

**FIG. 3.9: ESSEX BRIDGE, SHUGBOROUGH, STAFFS:** *This long, narrow structure dating from the 16th century illustrates how many packhorse bridges must have originally appeared, the low parapets having been designed to avoid contact with the animals' packs.*

## Road Bridges

FIG. 3.10: THORNBOROUGH BRIDGE, BUCKS: *With the loss of the monasteries and income from chantries after 1547, it usually fell upon local wealthy individuals to finance the repair of existing bridges or build new ones. This section from a plaque on Thornborough Bridge records work carried out upon its 14th-century structure in the 1660s. Shields on each side of the plaque contain the names of the two gentlemen who paid for the work.*

FIG. 3.12: STOPHAM, WEST SUSSEX: *A 15th/16th-century sandstone bridge over the River Arun. Its higher central arch, being segmental, is of a later date, probably raised to aid navigation in the early 19th century. The pointed cutwaters which cut back to a three-sided refuge can be found in other bridges from this period (see Fig. 3.2).*

FIG. 3.11: LANERCOST, CUMBRIA: *By the turn of the 18th century more large-span bridges were being built, like this example dating from 1724. This is a twin arch structure; the eastern one, shown here, is segmental and nearly 70 ft.*

# Tudor and Stuart Arched Bridges

**FIG. 3.13: BANGOR-IS-Y-COED, BRIDGE, CLWYD:** *A sandstone structure with five segmental arches crossing the River Dee. The parapets are made from single thin pieces of stone stood on end rather than built up in the conventional manner. The date stone recording its construction in 1658 is now eroded but once also included the date from the creation of the world, thought then to have been 5607 years earlier.*

**FIG. 3.14: WILTON BRIDGE, ROSS-ON-WYE, HEREFORD AND WORCS:** *This fine sandstone bridge of 1597 still has large piers, short spans, and ribbed arches like its medieval predecessors. Only the half-hexagonal refuges are characteristic of this date. The large sundial on the central pedestrian refuge is a Georgian addition.*

**FIG. 3.15: CORBRIDGE, NORTHUMBERLAND:** *Although the corbels on the underside of the parapet are Victorian and date to when the carriageway was widened, the rest of the structure dates from 1674. Nearby the remains of a stone-arched Roman bridge similar to that at Chesters has been found (Fig. 2.26).*

# CHAPTER 4

# *Georgian Arched Bridges*

**FIG. 4.1: PONTYPRIDD, GLAMORGAN:** *William Edwards' bridge over the River Taff was built in 1750. The three holes of increasing diameter at each end were to lighten the haunches and permit floodwater to pass through.*

William Edwards' family were farmers, yet he grew to be a mason who through endeavour and a touch of genius became a leading designer of bridges. He was also one not to give up! His first commission, in 1746, a bridge over the River Taff in Pontypridd, was damaged by trees hurled against it in a flood and collapsed. Yet he convinced the authorities that his new design of one large span 140 ft wide would not suffer the same fate and proceeded with the second structure, only for it also to

# Georgian Arched Bridges

fall into the waters, this time because of a fault in construction. It must be a credit to the man that he managed to get a third attempt, one which, benefitting from the lessons of the previous failures, still stands today. Edwards had realized that the large segmental arch had so much weight in its haunches (the lower side of the arch) that it was literally forcing the voissoirs at its fragile crest upwards. To counter this he made three holes through the spandrels, which served not only to relieve the weight acting upon the centre but would also reduce the pressure of flood water by allowing it to pass through. Edwards was one of a new breed that emerged in the 18th century, men who began as masons and millwrights but who, using new ideas, science, and mathematics, became architects of innovative, bold forms of bridges.

It was not until the turn of the 18th century that the condition of roads began to improve. Charging tolls for a section of road and passing this on to a trust who would be responsible for its maintenance had been tried before, but from 1700 the system was expanded. Acts of Parliament were passed (around ten a year at the beginning of the century and up to 50 by the end) to enable a local turnpike trust to be formed to control a section of main road, on average 20 to 30 miles long. (Money was collected at a gate, usually a spiked barrier which was turned out of the way of traffic, hence 'turnpike'.) Some trusts did little; others made a big difference, not only building new roads and reducing gradients on hills,

**FIG. 4.2: GREAT BARFORD, BEDS:** *This medieval bridge was widened and strengthened with the addition of the Gothic brick façade onto its stone superstructure in 1873. This started to pull away from the body and so tie bars with circular iron plates at the end were inserted 20 years later to hold it in place.*

FIG. 4.3: TYRINGHAM HALL, BUCKS: *This elegant single segmental arch was designed by leading architect Sir John Soane in 1793. It features arched niches and a radial arrangement of masonry in the spandrel wall. The twin-grooved decorative feature along the underside of the parapet is a favourite tool of Soane's.*

but also constructing bridges. Many classically-styled stone and brick structures were built by leading architects, while older, narrower structures were widened to permit the passage of carriages and carts side by side. Finance for these projects still came from the wealthy, many of whom would benefit from their construction, but increasingly major structures were adopted by counties and boroughs. They often took a more responsible attitude for their maintenance and built elegant structures to promote civic pride.

Although these improvements made travel a more reliable experience, by the turn of the 19th century further work was required to the actual structure of the roads to meet their growing use. Industry demanded heavier loads to be moved, and the well-to-do wanted to travel more frequently and for longer distances, inspired by visions of moors, mountains, and ruined buildings, as well as a frenetic social calendar. Better designs of carriages, improved breeds of horses, and numerous coaching inns reduced journey times but it took men like Metcalf, Macadam, and Telford to construct the properly drained and surfaced roads that made even greater speed possible.

At the same time the grand tour had opened the eyes of a new aristocratic youth to classical architecture and the work of Renaissance designers. They returned home with a burning desire to convert their pastoral estate into a sweeping vista from a lost Roman world, one which had been recreated on canvas by contemporary artists like Claude. Leading architects, converted to a taste for classical Roman orders and proportions, gave country houses a makeover, flooded valley bottoms to form serpentine lakes, and built elegant stone bridges to cross them.

## Types of Arches

The segmental arch was prominent, especially in the latter part of this period, when a greater understanding of the mathematics and properties of its design and the limits of the materials used enabled much wider spans to be built. Elliptical arches (a half oval shape with a smaller radius on the haunches than at the centre) were also popular, as they too could have large spans, but directed more of the load down into the ground than segmental arches. The ratio of the span of the arch to the width of the pier, which in medieval bridges had been 1:1, or 2:1 at best, was now, in the hands of leading engineers, reaching three or four times that figure. Semi-circular arches were also revived as a result of the renewed interest in Roman architecture, its pure form being seen as appropriate for a classical bridge, especially early in the period.

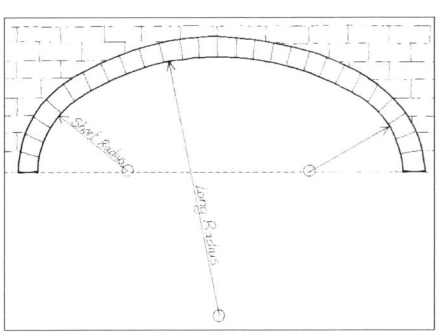

**FIG. 4.4:** *Diagram of an elliptical arch with two short radii on the springing line and a large one below.*

## Design and Construction

The most notable change to the design of arched bridges during the Georgian period was the increased length of the span. With the use of segmental and elliptical arches a river, which a hundred years previously might have been crossed by three or four arched structures, could now be spanned by one. The huge advantage of this was that piers did not have to be built in the water; the abutments on the river bank took all the outward forces, although as a result these had to be substantially stronger than in the past. In many situations, though, where there were large rivers on wide flood plains to be crossed and in towns and cities, high abutments were not practical. A steeply angled structure could be built (see Fig 4.1), but this could cause problems for carriages crossing and they were later lowered, as was the case in Pontypridd.

Wide rivers were therefore still crossed with multi-arched bridges, although, with wider spans, there were now fewer piers to block the flow of water, and they were also usually narrower. Where navigational limits and the height of approach roads permitted, some were built with a level deck, which was much more convenient for traffic.

Bridges were now wider than those built in previous centuries so that two carriages could pass, something which was rare before. Many also had pavements running the whole length and, as a result, the angled cutwaters of the piers no longer needed to be built up to the parapets to create refuges. On bridges of this period they are usually

# Bridges Explained: Viaducts & Aqueducts

**FIG. 4.5: WELSH BRIDGE, SHREWSBURY:** *The new, wider road bridges built from the mid-18th century had, as in this example, narrower piers and wider spans; cutwaters were angled back so that there were no refuges above but instead stone parapets with pavements behind for pedestrians.*

railway. Stone was still the material of choice for the finest bridges, and, with better transport, it could be sourced from further away. The blocks tended to be smooth finished, with fine joints between them (ashlar) and some, usually around the base, were roughly carved (rustication), a Classical detail derived from the bases of temples which appeared cut out of the underlying rock.

Construction was aided by improved machinery, including more efficient and heavier pile-driving machines and, by the turn of the 19th century, the first steam pumps which made it easier to keep working areas in the river dry.

**FIG. 4.6:** *Examples of fine cut ashlar masonry (top) and rusticated stonework (below). The latter was usually used on lower sections of buildings.*

tapered off and capped with a half pyramid or dome in line with the springing of the arch.

Another notable change was the vast increase in the number of bridges which needed to be built, not just for an improving road network, but also for new river navigations, canals, and, after 1825, railways. Although stone was still used, brick became very popular for these new types of bridges, usually where the base of the structure did not have to resist the scouring action of a river: for example, over a canal or

# GEORGIAN ARCHED BRIDGES

Cofferdams were widely used, but the invention of caissons enabled piers to be founded in trickier situations and deeper water (see page 16). Where the bed was soft, foundations were still established with timber piles and, in many, a grid of horizontal beams may have been laid over them, with stone rammed into the gaps to make a firm base for the masonry. Solid piers were built rather than rubble filled, and to spread the load the bases tended to step out a number of times unlike the single steps of medieval bridges. Above the arches the infill could still be rubble, but usually with some form of clay lining to keep water out of the arch joints.

As Edwards discovered with his second attempt at Pontypridd, the weight in the haunches of the bridge was a problem in these new large span arched bridges. His idea of creating tunnels through them was copied or adapted by others, while Thomas Telford, one of the most prolific bridge builders of his day, preferred to create a hollow structure with internal vertical walls supporting the deck above.

One of the problems with these new elegant structures with their thin piers and wider spans was the necessity of building all the arches at the same time in order to avoid the unsupported piers with segmental and elliptical arches being pushed over by the

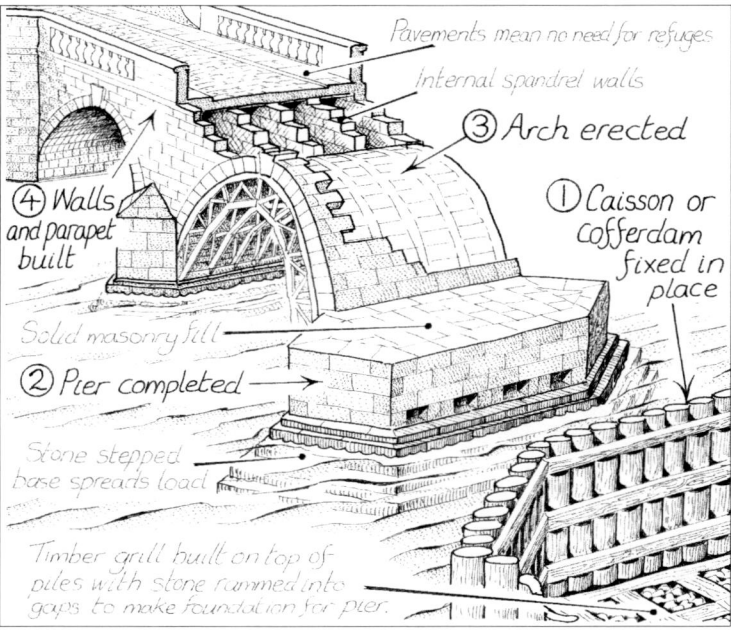

**FIG. 4.7:** *Diagram of how an 18th-century multi-arch bridge may have been constructed with elements from different bridges, although they may never have been used in a single project as shown here.*

# Bridges Explained: Viaducts & Aqueducts

outward forces. An improved understanding of trusses meant that traffic could pass beneath the wooden centring which supported the arch during construction.

Unlike earlier bridges, Georgian structures were finished off with classical decoration such as pilasters and statues. The formerly solid parapet walls were broken with open colonnades of turned stone and classical balusters with coping stones along the top, held together by iron straps. The keystone could be embellished with a raised rusticated finish or a pendant. The date and the name of the builder, or at least financier of the project, were often carved on a plaque proudly mounted at the crest of the bridge.

By the late 18th century leading figures sought a new identity for the

FIG. 4.8: CLARE BRIDGE, THE BACKS, CAMBRIDGE: *One of the earliest classically styled bridges in this country, built in around 1640. The central arch has a kink around the keystone owing to movement in the foundations, yet the structure is still secure.*

FIG. 4.9: *Classical-style details and decoration from late 18th-century bridges. Clockwise from top left: central plaque in a parapet; dentil horizontal course; statuary; triangular pediment and niche; prominent keystone; parapet with open balustrade.*

nation and the picturesque movement focused minds and eyes upon homegrown medieval structures, especially after the outbreak of the Napoleonic Wars brought the Grand Tours to a close. As a result, a new Gothic form of architecture appeared, at first heavily stylized and inaccurate, but, following the work of Pugin in the 1840s, it blossomed into a sharper more refined Victorian Gothic. Some bridges, notably on country estates, were built under the influence of this movement, often apparent just in the style of decoration but occasionally also in structural design.

FIG. 4.10: COSGROVE, NORTHANTS: *An example of a Gothic-style bridge with the distinctive flat pointed arch. This example was built around 1800 over the Grand Junction Canal.*

## EXAMPLES OF STONE-ARCHED BRIDGES FROM THIS PERIOD

## Ornamental Bridges

FIG. 4.11: BLENHEIM PALACE, WOODSTOCK, OXON: *This 100 ft segmental arch bridge was built by Sir John Vanburgh in 1708 for the Duke of Marlborough.*

FIG. 4.12: STOWE, BUCKS: *This Palladian-style bridge was based on a design by the Renaissance architect Andrea Palladio which was first used in this country at Wilton House, a few years before this example attributed to James Gibbs was built in around 1740. The far side was originally blocked off so that guests couldn't see the neighbouring village.*

# Bridges Explained: Viaducts & Aqueducts

**FIG. 4.13: CHATSWORTH HOUSE, DERBYS:** *The ornamental bridge designed by James Paine in 1759 completes the classic view of this most popular of English country houses. The statue is likely to have been re-sited here from elsewhere in the gardens.*

**FIG. 4.14: PULTENEY BRIDGE, BATH:** *This famous bridge was designed by Robert Adam and built in 1769. The central Venetian window, with shallow recessed arch above it, is characteristic of his refined work. The three segmental arches carry rows of buildings on both sides of the road (now used as shops). The bridge was financed by Sir William Pulteney when he was extending the River Avon navigation to his nearby estate.*

**FIG. 4.15: ILAM, STAFFS:** *Although medieval in style, a close inspection of this bridge dates it to the early and mid-19th century rebuilding of this estate village in the fashionable and more accurate Gothic style of this period. The cutwaters are rounded with rusticated piers, as on the classical structures of the day, and there are no refuges.*

# Georgian Arched Bridges

## Road Bridges

FIG. 4.16: SKERTON BRIDGE, LANCASTER: *Possibly the earliest completely level large road bridge, it was built in 1783 by Thomas Harrison and has elliptical arches and pedimented niches between.*

FIG. 4.18: ENGLISH BRIDGE, SHREWSBURY: *Designed by John Gwynn in 1768, complete with balustraded parapet, central plaque, carved keystone and dolphin statues on the cutwaters. As with his bridge at Atcham (Fig. 4.21), the steep angle of the roadway was a problem for traffic and the structure was partly rebuilt in 1926 to flatten it without destroying its original character.*

FIG. 4.17: OUSE BRIDGE, STONY STRATFORD, BUCKS: *Built in 1835, it has three shallow segmental arches, rounded cutwaters, and a dentil course running under the parapet.*

FIG. 4.19: MAIDENHEAD BRIDGE, BERKS: *This bridge across the River Thames was built in the 1770s by Sir Robert Taylor; it has large rusticated voissours and the tiniest of triangular cutwaters.*

**FIG. 4.20: GROSVENOR BRIDGE, CHESTER:** *When completed in 1833 this huge segmental arch was the largest of its kind in the world. Its designer, Thomas Harrison, decorated it in the more restrained neo-classical style of the day. Although nominally 200 ft wide, the arch is actually 230 ft, as it continues under the massive abutments.*

**FIG. 4.21: ATCHAM BRIDGE, SALOP:** *This graceful stone bridge was designed by John Gwynn and completed by him in 1776. The previous builder had been relieved after four years of attempting to build his timber grill and piles on the notoriously poor gravelly river bed. The concrete bridge in the background was built as its replacement in 1929.*

**FIG. 4.22: OVER BRIDGE, GLOS:** *Thomas Telford used a design of the great French bridge engineer Jean Rodolphe Perronet in this bridge completed in 1828. It features cornes des vaches (chamfered haunches with a lower elliptical and an upper segmental arch) for elegance and to help flood water pass beneath.*

# Georgian Arched Bridges

FIG. 4.23: BEWDLEY BRIDGE, WORCS: *Designed by Thomas Telford in 1799, it was built of sandstone, with three large segmental arches, pointed cutwaters, and classical decoration.*

## Canal and Railway Bridges

FIG. 4.24: TURNOVER BRIDGE, CONGLETON, CHESHIRE: *One of the most notable types of bridge on the canal network are turnover or snake bridges, which were designed so the horses towing the barges could swap banks without disconnecting the rope. The Macclesfield Canal has some of the finest examples.*

FIG. 4.25: LANCASTER CANAL, LANCASTER: *A stone humpback bridge typical of those built over canals in the late 18th-century boom. They can be of stone or brick with segmental or elliptical arches, sometimes with a single string course along the base of the parapet.*

# BRIDGES EXPLAINED: VIADUCTS & AQUEDUCTS

FIG. 4.26: DENFORD, STAFFS: *A simple stone bridge with a part segmental arch to give maximum clearance for boats.*

FIG. 4.27: MANIFOLD WAY, LITTON MILL, DERBYS: *Arched bridges over railways tended to be segmental or elliptical to create sufficient clearance for trains. Many main roads also pass over or under on an angle so that the railway didn't have to adjust its straight line. This, however, created awkward-to-build skew bridges (see Fig. 5.13).*

FIG. 4.28: MAIDENHEAD RAILWAY BRIDGE, BERKS: *This incredibly shallow elliptical brick arch was designed by Isambard Kingdon Brunel in 1837 for his Great Western railway. The 128 ft span is still regarded as the longest and flattest built. Many at the time thought it would collapse and so Brunel arranged a ceremonial knocking-out of the supporting centring without telling the crowd that he had slightly lowered the temporary structure some months before, so it was not supporting it anyway!*

CHAPTER 5

# Cast-Iron, Steel &
# Concrete Arched Bridges

**FIG. 5.1: IRONBRIDGE, SALOP:** *This iconic bridge of the Industrial Revolution is composed of five parallel iron-ribbed arches, which like a masonry arch, are all in compression, transferring the load of the deck down into the side abutments and the ground below. This was the first use of cast iron in this type of supporting role and its limits were unknown, yet, as it turned out, it wasn't the arch which gave problems but the stone abutments. The southern one (right in this picture) needed to be rebuilt further up the valley side because of movement in the foundations, hence the two iron-arched bridges, while the north abutment (left) still stands as the original southern one would have appeared. It, too, was found to be slowly slipping down the valleyside and had to be reinforced when the bridge was restored in the 1970s.*

# Bridges Explained: Viaducts & Aqueducts

By the second half of the 18th century the limits of masonry arches were being approached, yet the demand for bridges increased with the building of canals and, later, railways. The large costs in materials and construction of brick and stone structures led many to look for alternatives. One such arose in 1779, in a deep gorge above the tempestuous waters of the River Severn in Coalbrookdale. It was a project financed by local businessmen desperate for an immediate solution to the problem of an overworked ferry. The resulting ground-breaking revolution in engineering was used by the ironmasters in the group to promote their works to such an extent that the area which grew up around it became known simply as Ironbridge.

It was Thomas Pritchard, a leading architect, who encouraged the promoters of the bridge to use his design for a cast-iron arch between large stone abutments. His plans were initially based on a replication of the temporary timber centring used to support masonry arches during construction. Inspiration also came from new timber arched bridges on the Continent which used wooden ribs formed into arches to counter the various loads and pressures. At Coalbrookdale, however, cast iron, which was strong in compression as is needed for an arch, was in plentiful supply and of world-leading quality. After wavering somewhat, the promoters decided on Pritchard's recommendation, and his death in 1777 left the project in the hands of one of its most vigorous promoters, Abraham Darby III (the grandson of Abraham Darby who, in 1714, had first developed the use of coke in furnaces). It was this young ironmaster who became responsible for the bridge, although who exactly redesigned it with a semi-circular arch to create greater headroom for boats to pass beneath remains a mystery.

FIG. 5.2: IRONBRIDGE, SALOP: *Close-ups showing how the ribs were fixed together by wedges (top) and dovetail joints (bottom).*

# Cast-Iron, Steel & Concrete Arched Bridges

Once the stone abutments had been built up either side of the steep valley, the iron-ribbed structure, cast in sections at Darby's own Coalbrookdale works, was assembled on site in only three months, a fraction of the time that it would have taken to erect a masonry arch. All the pieces from which it is composed were pre-formed with joints, so that they dovetailed or mortised into each other just as in a timber structure. Iron wedges forced into holes pinned them in place; bolts or rivets, which hold most later structures together, were not used. The speed of work satisfied the promoter's desire for a quick fix. However, despite Darby's use of illustrations of it to advertise his works, the cost of the project crippled him financially and he never really benefited from it.

## DESIGN AND CONSTRUCTION

What the Ironbridge demonstrated to the thousands of visitors it attracted was that this seemingly fragile web of iron could successfully support the load from road traffic and that a large span could be assembled from prefabricated parts in a short time. Because it was considerably lighter than a masonry structure, the piers and abutments it rested upon did not have to be so substantial. This fact, coupled with the reducing costs of cast iron, would make a bridge of this material cheaper than one of stone. As cast iron worked best in compression, it was ideally suited to replace the masonry in an arch. From the late 18th century through to the middle of the 19th, cast-iron bridges were widely used for differing loads,

FIG. 5.3: ABINGDON, OXON: *Although this cast-iron bridge of 1824 has a shallow arch, it still transfers the load above it into the abutments. Although apparently similar, girder bridges with a flat deck are different as they form a self-supporting structure which rests on the sides and transfers the weight vertically.*

from foot to major road traffic, and can still be found today crossing canals and carrying road traffic.

The typical cast-iron bridge would be built between stone or brick abutments with an angled or stepped surface into which to fit the arched structure. The bridge itself was usually cast in a foundry, the name of which is often on the face of the ironwork. The arched ribs were made in matching halves, and in separate sections on larger bridges, to be fitted together on site. There would be from two to five ribs supporting the deck, fixed across by rods or bars. Early examples had concentric circles formed in the spandrels; later ones were often made

# Bridges Explained: Viaducts & Aqueducts

FIG. 5.4: *It was common for the foundry name and date to be cast on the structure.*

with a diagonal grid of ironwork. Many had decorative parapets, cast in the latest fashionable designs: simple geometric Regency forms in early examples; and Gothic emblems and pointed arches in the Victorian period.

FIG. 5.5: *Examples of an earlier spandrel design from 1797 with concentric circles and a more elaborate Gothic example from 1827.*

FIG. 5.6: *Examples of decorative cast-iron railings, from the simple examples of the early 19th century (top and middle) to the intricate work on the lower panel from 1878.*

## Steel-Arched Bridges

Although cast-iron bridges were widely used on the canal and road network, there were concerns when railways began to be built from the 1830s. Due to the immense weight of trains and the vibration caused by them, engineers tended to use brick and stone arches, and, later, newly developed trusses which utilized both wrought and cast iron for tension and compression pieces (see Chapter 6). The mass production of steel in the later 18th century gave engineers an alternative material which, being stronger in both tension and compression, permitted larger, more ambitious structures to be built, although it was not widely adopted for major projects until the 1890s.

If the gap to be spanned was too large for a truss bridge, especially over a large river, where piers mid-stream had to be avoided, a large single steel arch was often the best alternative. Rail and road bridges of this kind were assembled on site from individual bars and plates, bolted or riveted together, rather than large pre-cast iron sections as before. A disadvantage of carbon-based steel, however, is corrosion, and there are often complex layers of protective paints to counter this, most famously on the Forth Rail Bridge (see Chapter 7).

In a deep valley or gorge these large-span bridges would typically have the roadway on top of the arch. However, most crossings where a long bridge is required are on the flat, near the mouth of a river, where this is not suitable. One answer is to build the large arch and then suspend the roadway from it rather than rest it on top. The arch is still taking the principal load in the same way, but now the abutments and road leading up to it are at a low level. Sydney Harbour bridge is among the most famous examples of an arched bridge with a suspended roadway (Fig 5.7).

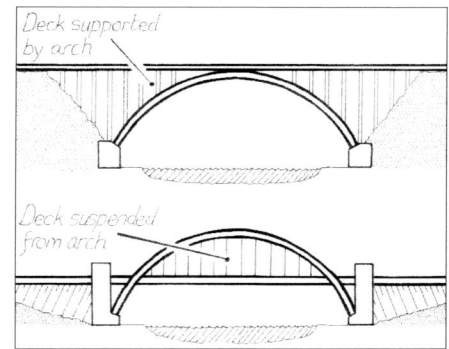

FIG. 5.7: *Diagram (top) showing the elevation of a large steel arch crossing a steep-sided gorge. It supports the deck in the conventional manner. This, however, is impractical in an open, flat valley and so the arch is constructed in a similar fashion but the deck is lowered to suit the terrain and is suspended on the section within the arch (bottom). This configuration is known as an arched bridge with suspended roadway and is used where heavy traffic loads are expected, as it is stronger and more rigid than a conventional suspension bridge (see Chapter 8).*

## Reinforced Concrete

At the same time as steel was being used for the first time on complete bridge structures, concrete appeared on the scene. At first it was used as a straight replacement for masonry on a number of arched railway viaducts (see Chapter 10). The development of reinforced concrete with, at first, wrought iron and, later, steel rods as a core around which it was poured transformed the type of structure which could be built.

**FIG. 5.8:** *A section of the reinforced concrete spandrel from the old Free Bridge at Coalbrookdale built in 1909. Note the tips of the metal rods protruding from the cut end.*

From the first decade of the 20th century arched bridges with open spandrels in a similar ribbed form to the earlier cast-iron bridges were built. Designs, however, were generally conservative and not as adventurous as some on the Continent, especially those of Swiss engineer Robert Maillart.

The real breakthrough in concrete came with pre-stressing – the tensioning of the steel rods or frames before the concrete set, which created a stronger material suitable for more daring shapes with less support (see Fig. 5.25). Modern concrete arches are usually made from box-shaped sections, which from the outside appear solid, and can be pre-cast to speed up production and reduce costs.

There are a great number of different ways in which bridges of iron, steel, or concrete can be built, and it can be difficult at first to discern whether the structure is an arch, beam or a truss. A true arch bridge is one in which the principal load bearing member is an arch in compression, the end of which meets the face of a low pier or abutment on the same plane (either directly or via a large hinge). This transfers the load along the arch and then out into the ground. At first glance many modern concrete and steel bridges have that appearance, but they are actually resting vertically upon the top of tall piers (often with rollers or cushioned gaps to allow for expansion, showing that there is no outward force). Any arch in this situation is forming part of a truss which has both tension and compression elements to keep it rigid. The next chapter will explain how beam and truss bridges work.

## EXAMPLES OF IRON, STEEL AND CONCRETE ARCHED BRIDGES

### Cast Iron

**FIG. 5.9: FRIARS GATE, DERBY:** *This railway bridge with elaborate cast-iron spandrels and balustrade was built in 1878. Even the underside is embellished with a lattice pattern.*

# Cast-Iron, Steel & Concrete Arched Bridges

**FIG. 5.10: COALPORT BRIDGE, SALOP:** *A similar scheme for a bridge over the Severn was put forward at the same time as Ironbridge but, built by a timber merchant rather than a foundry owner, it had two timber arches and a central pier. It was damaged in floods of 1795 and was replaced by three cast-iron ribs, the central one of which failed shortly after, resulting in a major rebuild with five ribs and a new cast-iron deck and balustrade in 1818.*

**FIG. 5.11: CANTLOP BRIDGE, SALOP:** *The sole surviving cast-iron bridge built by Thomas Telford in his role as county surveyor. Anthony Blackwall, the county's bridge engineer in the 1970s, commissioned a new concrete structure adjacent to it when hairline cracks in the ribs had been reported by a keen-eyed passer-by. However, it later transpired that the cracks reported were, in fact, the gaps that had been left by Telford for expansion and that the replacement bridge was probably unnecessary.*

**FIG. 5.12: TICKFORD BRIDGE, NEWPORT PAGNELL, BUCKS:** *In the early days of cast-iron bridges an alternative method of building them was developed by using shorter cast pieces like the voussoirs of a masonry bridge, which were fixed together with wrought-iron straps. After a number of failures, this type of construction was dropped, but this example from 1810 has stood the test of time. The circle pattern inserted into the spandrels is very typical of early cast-iron bridges from this period.*

# BRIDGES EXPLAINED: VIADUCTS & AQUEDUCTS

**FIG. 5.13: CASTLEFIELD, MANCHESTER:** *This impressive cast-iron bridge over the Rochdale Canal was completed in 1849. Like many railway bridges, it was built on the skew so that the tracks would not have to be re-aligned; each rib was set into individual stone and brick skewbacks (abutments). The spandrels have a Gothic narrow lancet design, which is typical of this period.*

**FIG. 5.14: EATON HALL, CHESHIRE:** *A large ornate cast-iron bridge of 1824, designed by Thomas Telford. It crosses the River Dee on one of the Duke of Westminster's estate roads.*

**FIG. 5.15** *(below)*: *Cast-iron footbridges can still be found on a large number of canals.*

**FIG. 5.16: BETWS-Y-COED, GWYNEDD:** *Built by Thomas Telford in 1815, it bears the inscription 'This arch was constructed in the same year as the Battle of Waterloo'. The spandrels contain castings of a leek, a rose, a thistle and a shamrock.*

# Cast-Iron, Steel & Concrete Arched Bridges

**FIG. 5.17: CHEPSTOW BRIDGE, GWENT:** *A large cast-iron bridge designed by John Rennie and opened in 1816. Rather than a single large span, it has five cast-iron ribs resting on stone piers with rounded cutwaters. Like Ironbridge and Coalport Bridge, it has a decorative panel in the centre of the railings.*

**FIG. 5.18: ALBERT EDWARD BRIDGE, SALOP:** *An unusually late example of a cast-iron railway bridge with a 200 ft span. The nine sections which make each of the four ribs are bolted together. It was designed by John Fowler, who went on to design the Forth Rail Bridge (see Chapter 7).*

**FIG. 5.19: KINGSLAND TOLL BRIDGE, SHREWSBURY:** *This early example of a metal arch bridge with suspended roadway was built in 1882 by the Cleveland Bridge and Engineering Company, who are still of renown today. Its 212 ft span carries the road sufficiently high above the River Severn to avoid problems with flooding.*

# Bridges Explained: Viaducts & Aqueducts

**FIG. 5.20: SILVER JUBILEE BRIDGE, RUNCORN, CHESHIRE:** *Opened in 1961, this huge steel arch with suspended roadway was designed by Mott, Hay, and Anderson, the most prolific bridge-design company in the 20th century. It is the largest of its type in the country (over 1000 ft) and about two thirds the size of Sydney Harbour Bridge. Its constant repainting programme takes about five years.*

**FIG. 5.21:** *Details from Tyne Bridge: the huge steel hinge which links the arch to the abutment (left); and the Art Deco doorway in the base of the tower (right). This led to lifts, which were originally intended to serve warehouses inside the superstructure, but were never finished.*

**FIG. 5.22: TYNE BRIDGE, NEWCASTLE:** *Another Mott, Hay, and Anderson design, this one over the River Tyne, with a span of 531 ft. It was based on the planned Sydney Harbour Bridge and the earlier Hell Gate Bridge, New York, and was completed in 1928. When the arch was first being erected without the roadway some locals complained that it would be too steep to get their horses up!*

## Cast-Iron, Steel & Concrete Arched Bridges

**FIG. 5.23: NEW ATCHAM BRIDGE, SALOP:** *A reinforced concrete arched bridge completed in 1929 to bypass John Gwynn's old Atcham Bridge (see Fig 4.21). To respect its older neighbour, it was styled with classical piers, balustrading, and a central plaque, a rather odd but appropriate mix.*

**FIG. 5.24:** *Concrete became the common building material for road bridges in the 1930s. Many from the inter-war years, like this example from Lincolnshire, were decorated in Art Deco style with plain, massed blocks and streamlined horizontal grooves. Small pebbles and gravel were often mixed in to give this modern material a more vernacular finish.*

**FIG. 5.25: SCAMMONDEN BRIDGE, WEST YORKS:** *This reinforced concrete arched bridge over the M62, high on the Pennines, was built in the late 1960s and, at 410 ft, was the largest single span bridge of its type in the world at the time. Its twin box section arch and vertical spandrel walls were designed carefully to reduce disturbance to the air flow through the cutting, which could cause snow drifts.*

## CHAPTER 6

# *Beam, Truss and Girder Bridges*

**FIG. 6.1: ROYAL ALBERT BRIDGE, SALTASH, CORNWALL:** *This most challenging of Brunel's railway projects resulted in one of his most ingenious designs, a complex yet graceful pair of 460 ft spans which met all the conditions set him. Although they combine large arches with a suspended deck, the resulting structures are self-supporting and as such are trusses, probably the most complex version of a type which originated as the simple beam.*

Brunel's railway engineering genius was challenged on many occasions but nowhere more so than at the crossing of the River Tamar at Saltash on the border between Devon and Cornwall. This large waterway was steep sided and deep, the very reason why the naval docks were built further upstream, and it was the arrival of demands from the Admiralty when

# Beam, Truss and Girder Bridges

planning was well underway which would cause Brunel the headache. The Admiralty insisted not only that there was sufficient height for sailing ships to pass beneath but also that there was clear width for their wide masts. On top of this, the waterway was to be clear at all times during construction for them to pass!

One single arch was out of the question because of the limitations of materials and because the temporary scaffolding needed to build it would block the river. Suspension bridges were discounted, as they were proving unsuitable for railways (see Chapter 8), and a conventional girder bridge would have too many piers, causing obstructions to this tidal route. Brunel's solution was to span the river with two huge 460 ft spans supported on two outer piers and one central pier. Due to the height required and the tall narrow piers which would be used to carry the railway over the river, the spans would have to be self-supporting and exert no lateral force, all the weight and load acting vertically. The final design was one of the most complex examples of a truss, a type of bridge which acts in the same way as the simplest form, the beam.

## BEAM AND GIRDER BRIDGES

The easiest way to bridge a gap is to use a horizontal beam resting on each side. The force from its own weight and that of the load crossing it is carried vertically on the supports at each end, with little or no outward horizontal pressure such as there is with an arch. In a bridge this means the piers and abutments can be less substantial and, as a result, also taller.

The limitations of a beam are determined by the material and how it acts under a load. If you imagine crossing a plank of wood suspended between two stacks of bricks, it will bend, depending on your own weight and that of the timber itself. The bend will be at its greatest when you reach the middle. If you were to look at it from the side at this point, the displacement would be running horizontally along the centre of the piece of wood, so that the section of the timber above the line is squashed up (in compression) and the section below is being stretched (in tension). A beam therefore is best made from a material which is both strong in compression and tension (Fig 6.2).

Stone works well in compression but is brittle in tension and so a slab thin enough to make a practical deck will

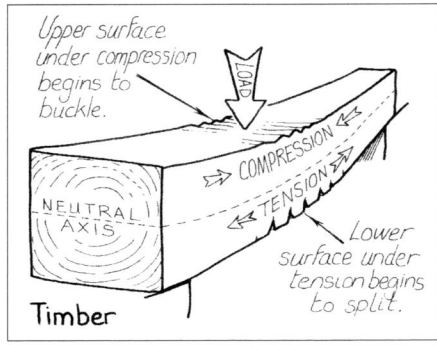

FIG. 6.2: *A diagram showing how a timber beam reacts to a load being applied along the upper surface, as when someone crosses one over a stream.*

fail if the span is anything more than around 10 to 20 ft. Wood is far more suitable, being flexible and far easier to source, move to site, and work with. This is why most beam bridges from prehistory through to the 18th century were built of wood.

There are three forms of beam bridge which are known to have been built from ancient times through into the modern age. The most simple form and one which survives today, albeit rebuilt over the centuries, is the clapper bridge. In areas where stone was readily available large blocks were positioned across the river or stream to form low piers, and slabs were then laid across them. Good examples survive in Devon, although they can be found in most upland areas, including the Peak District and the Lake District. The problem is that, with no dating evidence on the structure itself, it is impossible to know how old they are, although documents indicate that many are on the site of an ancient crossing.

FIG. 6.4: POSTBRIDGE CLAPPER BRIDGE, DEVON: *One of the few surviving clapper bridges which is likely to be of an ancient date, although its suggested Bronze Age origins are impossible to prove.*

FIG. 6.3: YOULGREAVE, DERBYS: *A simple clapper bridge resting on an assortment of different stones, its current form probably of no great antiquity.*

Surprisingly, timber beam bridges are now known to have existed from antiquity and with dendrochronology (dating wood from known patterns of tree rings) these can date back as far as 3000 BC. The earliest examples found are the prehistoric trackways preserved in peat, most notably the Sweet Track in the Somerset Levels. These are formed from pairs of vertical or diagonal trunks piled into the marshy ground with horizontal planks placed between them. Bridges dating from the Bronze and Iron Ages, with vertical timber piles and horizontal beams supporting a planked deck, have also been uncovered.

The Romans were probably the first to combine these materials into a third form, consisting of stone piers built up from the river bed and a timber deck spanning between them, possibly with

# Beam, Truss and Girder Bridges

**FIG. 6.5: ETON ROWING LAKE, BERKS:** *A reconstruction of an Iron Age bridge based upon one found during excavations by Oxford Archaeology at the site of the new Eton Rowing Lake in the late 1990s. Numerous vertical lines of round oak timbers piled into the river bed were found, dating from as far back as 1300 to 1400 BC. Lying around one example were found what were assumed to be the horizontal beams and wooden squared planks which made the upper part of it. It was surmised that a projecting tenon on the top of the vertical post held the side beams (which were found with mortice holes) in place, although it is not clear how the smaller planks which lay across these to make the walkway were fixed.*

**FIG. 6.6: PIERCEBRIDGE, NORTH YORKS:** *The excavated remains of an abutment from a Roman bridge believed to date from the 2nd century AD. This and the piers would have been constructed of masonry, with an unknown form of timber deck. Reconstructions have used a complex truss structure from Romania featured on Trajan's Column in Rome. The actual bridge may have been far more simple than this. The notches on top of the first course of masonry may have been for temporary or permanent supports for the deck. The stone surface in the foreground is paving to prevent scouring and to keep the current flowing smoothly.*

additional support from diagonal struts resting on corbels or beams set in the stonework. Most of their major bridges probably took this form, and it is likely to have continued in use into and throughout the medieval period. This kind, like the other two types of simple beam bridge, can still be found today on minor crossings.

## Metal and Concrete Beam Bridges

With the wider availability and improved quality of cast iron in the later 18th century it became suitable for use in simple beam bridges. Rather than a solid block it was formed into a vertical thin plate strong enough to resist shearing with a strengthening horizontal piece running along the top and bottom (to form a 'H' or 'I' shape in cross section). However, because cast iron is weak in tension, the beam could be thickened or widened at the bottom where the tensile forces were at their greatest to provide strength where it was needed. These metal girders could be simply laid parallel to each other on top of abutments either side of a crossing to support a deck above in the same way as a timber structure.

Steel is equally strong in tension and compression so beams in this material usually have similar sized top and bottom flanges (you can often tell earlier cast-iron girders from this thickened bottom flange). Concrete is also used as a solid beam in short-spanned bridges and being weaker in tension it is again strengthened on its lower part but this is done more often by having extra steel reinforcing rods or wires in this section than by any external thickening (note the extra rods sticking out of the upper part of the bridge in Fig 5.8).

FIG. 6.8: MARPLE, CHESHIRE: *A plate girder railway bridge with both parapets and supporting beams beneath formed in this way.*

Metal girder bridges became an important form for railways from the 1830s as they could be quickly and cheaply formed and yet were rigid and strong enough to take the load of a train. However, the brittle nature of cast iron resulted in a number of catastrophic failures. A notable example befell Robert Stephenson at Chester in 1847 when four people died when a girder on the new bridge over the River Dee broke as a train was crossing. Most engineers preferred to use wrought iron which was stronger in tension and more reliable either in combination with cast iron or on its own, this in turn being replaced by steel towards the end of the century.

Most small crossings were built in the form of a plate girder bridge. These

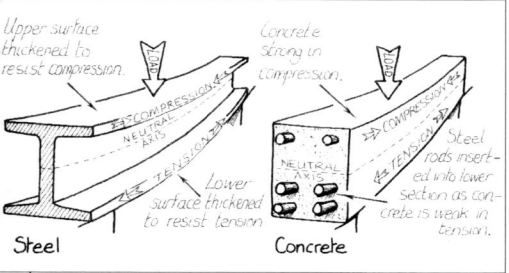

FIG. 6.7: *Diagram of steel and concrete beams supported at each end and under pressure from a load in the middle. The dashed line in the middle is the centre of the displacement, the neutral axis.*

# Beam, Truss and Girder Bridges

use larger thin metal panels with a bolted or riveted frame of thicker flanges to make a rigid supporting structure either just up the sides or underneath as well.

## Box Girder Bridges

Simple girders were, however, not strong enough for larger spans across rivers or valleys, and engineers used calculation and experimentation to find other forms which could retain their rigidity and strength over these greater lengths. One type was the box girder in which, rather than having a metal base and sides, the railway line is completely boxed in with a fourth plate above. These structures, however, are far more complicated than first appears.

FIG. 6.10: *A section of the original box girder used on the Britannia Bridge.*

FIG. 6.9: BRITANNIA BRIDGE, GWYNEDD: *As a safety precaution in the original plan, suspension chains were to be hung from the towers to support the box girders. In the event they were never fitted but the holes at the top of the towers for them can still be seen today.*

The most famous example of this form was the Britannia Bridge which carries the Chester to Holyhead railway across the Menai Straits built by Robert Stephenson and completed in 1850. The problem he faced was the same as Brunel was to face at Saltash, the Admiralty insisted that this important sea lane was not obstructed by temporary scaffolding and had a high, even clearance so that sailing ships could pass beneath. His advantage over Brunel was that the chosen site had an island midstream (Britannia Island) which would provide an invaluable centre support.

After consulting leading engineers and manufacturers, Stephenson decided upon two parallel rectangular tubes built from wrought-iron panels, riveted together. For additional strength at the bottom and especially the top which was under compression (where this material was not so strong) he built cells out of girders, 8 along the top and 6 along the bottom (see Fig 6.10). These huge tubes were constructed on the shore and then floated out into position where they were raised up within gaps in the piers, 6 ft at a time, by hydraulic presses positioned on top of them and wood jacked in case these failed. After each short lift, masonry was built up beneath in the gaps to complete the piers. As only one side was being worked on at a time the other was open to shipping, meeting the Admiralty's requirements.

FIG. 6.12: CONWY TUBULAR BRIDGE, GWYNEDD: *A forerunner of the Britannia Bridge along the same Chester to Holyhead railway, it was the earliest tubular bridge (completed in 1848), in which the trains run through the box-girder. The portals were designed to match the adjacent castle.*

FIG. 6.11: BRITANNIA BRIDGE, GWYNEDD: *After the fire in 1970, the spans were rebuilt as steel arches supporting the railway, with a road deck above. The towers are original, and the famous lions which flank the entrances remain, if currently out of view.*

The bridge was a success and lasted until a tragic fire in 1970 buckled the box girders so much they had to be replaced by a steel arch structure carrying both rail and road. A slightly earlier box girder bridge built by Stephenson for the same line survives in almost original form at Conwy alongside Telford's suspension bridge (see Chapter 8) from which it can be closely examined.

Although this form of bridge was rare, the idea of using a box form for strength has now become common. Steel and reinforced concrete box girders are widely used in large bridges. Although they may appear solid from the outside they are hollow to keep the

# Beam, Truss and Girder Bridges

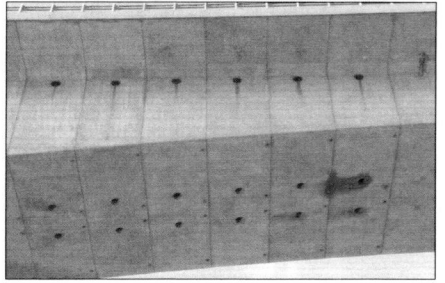

FIG. 6.13: *A modern concrete road bridge over the Tyne in Newcastle with a close-up below of the box girder sections it is made up from. Although they have an arched underside, the thin piers on which it rests show that it is still principally a hollow beam with all the load transferred vertically downwards.*

weight down (the arch in Fig 5.25 is made from twin hollow concrete boxes). Concrete box girders can be formed into various shapes, often with an arched underside, and supported by numerous tall piers carrying main roads and motorways across some of our largest rivers.

## TRUSS BRIDGES

Stephenson's wrought-iron box girder turned out to be an expensive option, designed as it was for a unique circumstance. Most railway bridges used another type of structure, a truss, which could extend beyond the length of a beam. In its simplest form the truss could be described as a beam with sections cut out so that the structural elements remain but the overall dead weight is reduced in order that it can stretch further. Most importantly, the elements are arranged in triangles, as this is the only geometric shape which will not distort: a circle can be squashed into an oval and a square displaced to form a rhombus, but a triangle always has a length which will resist the movement on the other sides. This was appreciated in Europe as early as the Renaissance, and leading architects like Palladio drew examples of bridges using trusses. Although a number of small-scale bridges were built, it was not until the coming of the railways in the early 19th century that engineers began to use the design widely.

It was in America where most development took place as, with a shortage of materials and skills, trussed bridges provided cheap, easy to erect structures which could be built using the plentiful supply of timber from the forests through which the railroads passed. A wide number of different patterns were drawn up, named after their designers. The Bolman, Fink, Pratt, and Howe trusses comprised diagonal tension and vertical and horizontal compression members.

In Britain trusses made with wrought iron for the tension parts and cast iron for the compression elements were used on railways. The most common early

# Bridges Explained: Viaducts & Aqueducts

FIG. 6.14: RUNCORN, CHESHIRE: *Lattice girder railway bridge with decorative castellated turrets. Completed in 1868, it crosses the Mersey.*

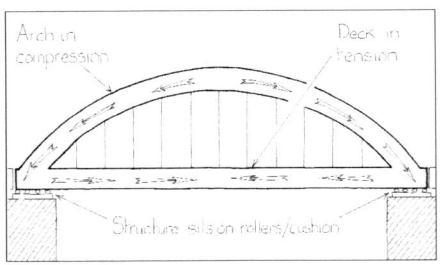

FIG. 6.16: *Diagram of a tied arch. The arch in compression is tied back by the horizontal tension member, in this case the deck. As they counter each other, there is no outward thrust so the structure can simply be positioned on top of piers with cushions or rollers to allow it to expand and contract.*

form was the lattice girder. Later in the century as steel, which was strong in both tension and compression, became widely available, and stresses in these types of structures were better understood, simple truss bridges like the Warren girder were widely used, either with the steel framework above the deck or below.

FIG. 6.15: *A Warren girder, a simplified form of truss which was popular for steel railway and road bridges.*

## Bowstring Girder or Tied Arch Bridges

Often confused with arches with suspended roadways (see Chapter 5), these structures were an early answer to the problem of spanning large rivers without imposing lateral thrust on high piers. They are in fact a truss, as the arch, rather than transferring the load outwards, is held back by a horizontal tension member, thus tying it in place so that all the dead and live load is exerted downwards onto the piers. Further members are used to hold it together and add rigidity.

The earliest notable example is the High Level Bridge in Newcastle, built

**FIG. 6.17: HIGH LEVEL BRIDGE, NEWCASTLE:** *The first major bridge to combine wrought and cast iron. It consists of 125 ft tied arch spans, each with four cast-iron arch ribs, with the railway above supported on cast-iron columns, and the roadway below hung on wrought-iron rods from the arches.*

the central pier, at Saltash there was none. To make matters worse the riverbed sloped and was up to 80 ft deep. Brunel had to build one of the first examples of a pneumatic caisson, a huge wrought-iron cylinder deeper than the river itself which was floated out and sunk into position. Pressurized air forced the mud and water out at the bottom so that it could be excavated until bedrock was reached (see Chapter 1). The masonry was built up inside until it reached above high water level and then, to reduce the load bearing down upon it, hollow iron columns were built on top of this with a braced gap between. This space was needed to allow each of the tied arched trusses (known in this case as fish-bellied girders because of their shape) to be lifted into position. These two huge structures weighing over 1,000 tons

by Robert Stephenson and opened in 1849. Here cast-iron girders and wrought-iron tension rods were used in combination to support a railway along the top and a roadway within the tied arch structure supported on stone piers. This ingenious system was a remarkable achievement but the spans were only of 125 ft. When Brunel took the Cornish Railway across the Tamar at Saltash, Plymouth, a few years later, he faced the task of creating trusses nearly four times the length.

Like Stephenson when he bridged the Menai Straits, Brunel faced a problem in meeting the Admiralty's requirements in crossing the Tamar, but, whereas the Britannia Bridge had an island midstream on which to found

**FIG. 6.18: ROYAL ALBERT BRIDGE, SALTASH, CORNWALL:** *One of the few bridges to have the engineer's name displayed prominently, a reflection of Brunel's high standing.*

# Bridges Explained: Viaducts & Aqueducts

each had been fully assembled on land and then floated out on a pontoon to be raised into position by means of hydraulic presses. As at Britannia Bridge, they were positioned one at a time so that half the river remained open for shipping at all times.

After six years of building, the Royal Albert Bridge was opened by Prince Albert on 3rd May 1859. Brunel, however, was not there; he was seriously ill, and only crossed his great project in a wagon a short while later, before his death on 15th September 1859. As it was to turn out, the affection and respect he commanded meant this was not his last work, as will be seen in Chapter 8.

For most of the railway age, tied arches and lattice trusses were the preferred choice for major bridges, but the failure of the Tay Bridge in 1879 – lattice trusses were used here, although not a direct cause of the collapse – was to result in engineers looking at an alternative design. This new form, the cantilever bridge, could potentially employ even longer spans and, hence, fewer piers, yet remain rigid enough for trains to safely pass over.

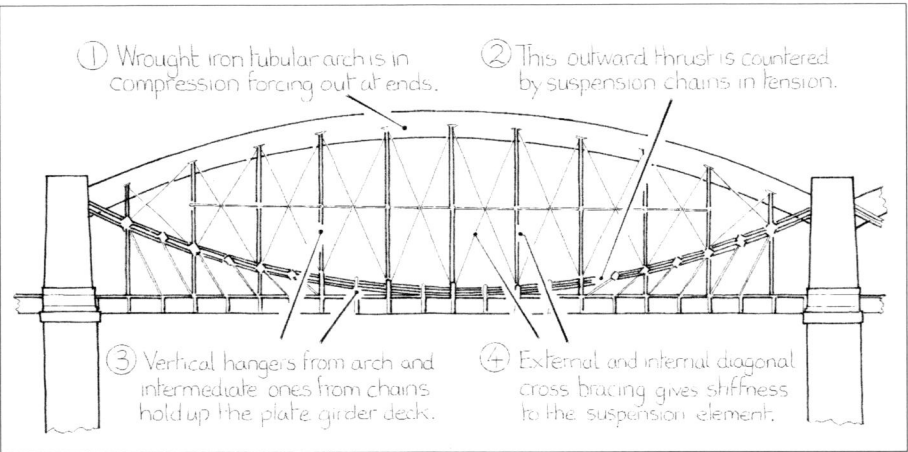

**FIG 6.19:** *A diagram showing the fish-bellied or tied arch used by Brunel on the Royal Albert Bridge (not to scale) and how the different parts work together. Although these form a truss, the deck is suspended; the only example where this happens on a British main line railway.*

# Chapter 7

# *Cantilever Bridges*

**FIG. 7.1: FORTH RAIL BRIDGE, LOTHIAN:** *This most famous of British bridges, 5,350 ft long (8,296 ft, including approach viaducts), was the first major structure built in steel and was also the first built here using the cantilever principle. It was completed in 1890.*

## The Cantilever Principle

Unlike a simple supported beam, which is held up at both ends, a cantilever is a beam which is fixed at one end so that the horizontal projecting part can support a downward load. In this type, major considerations are the strength of the fixing and how deep it is set, and the material of the beam and how far from the fulcrum the load will be. Staircases from the 18th century have often been fitted this way, seeming to float up without any vertical newel posts for support, when in fact the treads are held by being set deep into the wall. An alternative way of cantilevering a beam is to project it out on a bracket with an angled arm fixed to its underside and set into the wall. Shelves and balconies work in this way (Fig 7.2.A).

In cantilever bridges the same principle applies: the deck is supported from beneath by a vertical pier or angled arm. The arms can be built out from either side to join in the middle, with just a fine central gap (left to allow for expansion) to reveal the method of construction. (Fig 7.2.B) This form, however, offers little in extra length over a normal beam bridge. The real advantage is that the two cantilevers can be moved further apart and a central section inserted that rests on their ends, making a much larger span than could be produced in a simple beam bridge (Fig 7.2.C). Unlike an arch bridge, a cantilever bridge does not need centring to support it during construction; as long as the structure is fixed or balanced, the cantilever arms and central section can be built with minimal disruption to traffic passing below, which was important if bridging a canal, river, or estuary. The design also retains the beam or truss bridge's advantage of firmness. Although suspension bridges could do a similar job, they were not suitable to hold the single displaced weight of a passing train.

Cantilever bridges were built in China from at least the 4th century and known in Europe probably as a result of the Crusades in the Middle Ages. In their simplest forms they were timber structures with overlapping beams projecting from either side. Cantilever structures, however, have both strong compressive and tensile forces and are best built with a material strong in both. It was not until the mass production of steel and its subsequent drop in price in the 1870s that this metal, which was stronger than iron in both compression and tension, could be considered. The notable American engineer James Eads had been the first to demonstrate its suitability for major bridges with his huge arched structure over the Mississippi opened in 1874.

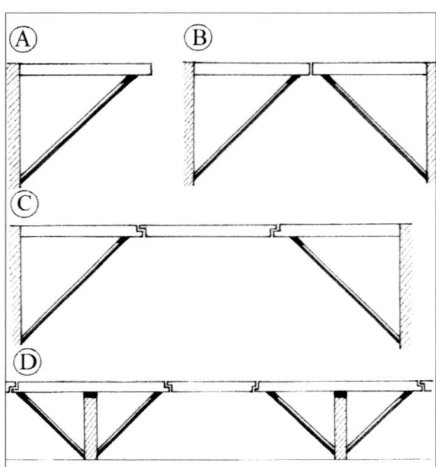

**FIG. 7.2:** *Diagram showing progression from a simple cantilever to a large bridge. (A) shows a shelf which, if matched by another, can span a gap, (B). If these are pulled apart, then a suspended section can be inserted, making a larger span, (C). If a cantilever of equal size and weight is built out from the other side of a pier, they can balance each other, making an even longer bridge, (D), with very few piers compared with a simply supported bridge.*

Despite that, there was a reluctance to accept it on these shores.

## The Forth Rail Bridge

The Forth Rail Bridge is one of the most famous engineering landmarks in the world. The choice of its cantilever form was a direct result of the Tay Bridge disaster (see Chapter 1). Sir Thomas Bouch had already laid the foundation stone for his planned suspension bridge over the Firth of Forth when the Tay Bridge collapsed on 28th December 1879. Despite the huge scale of the project and cost in money and lives, the bridge over the Tay had in its short life proven profitable, and the North British Railway Company, keen to press ahead with its rebuilding, appointed Benjamin Baker and John Fowler, both experienced engineers. Their replacement structure was more substantial but fundamentally the same with a line of piers supporting lattice girder trusses.

Even as building began here the pair were already turning their attention to the crossing of the Firth of Forth, which had been held up after the Tay disaster. Although the distance was slightly shorter, the water was much deeper and it would not be possible to build a series of piers and trusses all the way across; the idea of a suspension bridge was also dismissed. Fears of financiers (the Forth Rail Bridge was jointly funded by four railway companies, including the North British) and the general public in the wake of the Tay disaster meant the new structure would have to be strong enough to resist winds far in excess of what was likely. At the same time it had to be rigid enough for the concentrated load of passing trains and yet have huge spans, as only one pier would be possible, in the centre, where there was an outcrop of rock. The bridge would also have to be built without the aid of scaffolding or centring because of the deep water and in order to avoid obstructing shipping.

The answer may have already been known to Fowler, who some years before had mentioned the idea, but it was first demonstrated to be practical in Germany in 1867, with Heinrich Gerber's Hassfurt Bridge. This, however, had a central span of only 124 ft. The new Forth Rail Bridge had an 8,000 ft crossing with main spans of over 1,700 ft each, making the design which Baker and Fowler chose the first major bridge in the world to use the cantilever principle and the first all in steel.

The design comprised three large balanced cantilevers, linked to each other by a pair of 350 ft spans and to the shore by a north and south approach viaduct formed from conventional trusses and piers. The great towers at the centre of the balanced cantilevers each stood upon four piers set into the river bed. They were built using pneumatic caissons (see Chapter 1). Upon these were built the steel tubes which formed the centre of the tower, held in position by cross-bracing, and then, from them, the two cantilevers were progressively built out by the same amount each side to keep them in balance. The lower arm or strut was a tube, which would take the compressive force, and the upper one a

# BRIDGES EXPLAINED: VIADUCTS & AQUEDUCTS

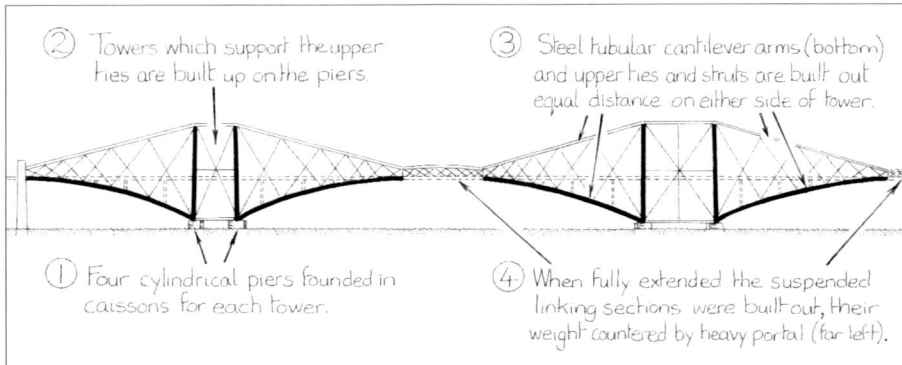

**FIG. 7.3:** *From the shore the structure of the Forth Rail Bridge looks a bewildering arrangement of tubes and girders, but its structure can be broken down into simple stages as in this diagram. It shows two of the three spans and the basic process of construction from 1-4. This bridge is in principle the same as that in Fig. 7.2.D, with the main supporting cantilever arms and the lower arched pieces outlined in black, as they are in compression. However, due to the immense size and the environmental conditions in the Firth, extra support was needed, which is why the towers (also in compression) are built so high; from them lattice girder ties (in tension) run down to the ends. The cross-bracing between is used to support the arms during construction and to make them rigid. This robust bridge is the only one in the country over which trains can travel at full speed.*

**FIG. 7.4: FORTH RAIL BRIDGE, LOTHIAN:** *A view from the north shore during repainting in 2005.*

lattice girder tie, which would hold it in place under tension, with diagonal braces between them. When they had reached their full length the outer arms of the north and south spans were fixed to the end portals of the approaches, large structures which acted as 1,000-ton counterweights for the suspended central sections. These were built out between the three spans and were fixed to the cantilever arms with huge pins the size of a man.

Although the Forth Rail Bridge demonstrated the strength and suitability of steel for construction, it also showed its weakness – corrosion;

steel is far more prone to rusting than iron. The regular protective painting of the Forth Bridge is essential for its survival, and, because of the huge surface area, painting famously has to begin from one end almost as soon as the last coat has dried on the other! In the future, however, this may no longer be the case, as a new system is being applied using a zinc primer, glass flake epoxy, and a polyurethane gloss red, as used on oil rigs, and this should have a 20 to 30 year life.

**FIG. 7.5: WARBURTON BRIDGE, CHESHIRE:** *Built in 1894 over the Manchester Ship Canal, this steel structure has two balanced cantilevers with a central suspended section between them (as shown in Fig. 7.2.D). The lower photo shows the expansion gap between the sections.*

Following the success of the Forth Rail Bridge other structures were built using the same principle and material, including two examples across the Manchester Ship Canal, which was being built in the 1890s. However, the end of the railway age was already in sight with the dawn of the internal combustion engine. For this new transport system, which grew during the 20th century, reinforced concrete and steel arches and simply supported beam structures seem to have been preferred. However, cantilever bridges have proven very popular in the last 50 years and, in one particular location, motorways, continue to be so today.

## Motorway Bridges

Perhaps the most underrated form of bridge is that which we pass under in our millions every day. Motorway and major road concrete bridges may seem monotonous to the speeding motorist, and yet they are far from standard, each requiring an individual design suitable for the location and each displaying creativity, engineering flair, and beautiful form in many cases. They are also bigger than we perhaps appreciate. If asked which was the largest single-span concrete arch bridge in Europe, few would probably say that it the one that carries the B6114 over the M62 at Scammonden (see Fig 5.25). Even a footbridge has to be carried across three lanes on either side, a central barrier, two hard shoulders, and often cutting sides too.

To span these very large gaps economically and create fewer obstructions in the form of piers, the

# Bridges Explained: Viaducts & Aqueducts

FIG. 7.6: CASTLE BRIDGE, SHREWSBURY: *Built in 1951, this was the first pre-stressed concrete counter-balanced bridge in the country. The central section is supported by the two concrete box girders balanced over piers and fixed to the far bank. To counter the tension in the upper section of the cantilever box girders and the lower part of the central section (acting like a beam), high tensile cables were run through the inside of the whole bridge, tightened, and then fixed. It is the pre-stressing which makes these simple obstruction-free structures possible.*

cantilever principle was revived. If you take a closer look at a bridge over a motorway, you may notice a stepped gap in the deck, marking the central supported section, with the cantilevered sides balanced on piers or supported by angled columns.

When these major routes came to a large river or estuary it was not the large steel cantilever like the Forth Rail Bridge which was built. As cars in these situations do not form a concentrated load like a train, they do not need such a rigid structure and so this made another type of bridge, the suspension bridge, a more economical and suitable choice.

FIG. 7.7 *(left)*: *A reinforced concrete cantilever motorway bridge. The angled columns support the deck on each side, forming cantilevers with a central section suspended between them.*

# CHAPTER 8

# Suspension Bridges

**FIG. 8.1: MENAI SUSPENSION BRIDGE, GWYNEDD:** *Thomas Telford's London to Holyhead road faced its greatest obstacle at the Menai Straits, where restrictions imposed by the Admiralty effectively forced him to build the first major suspension bridge in the country. With a 580 ft span and towers over 150 ft high, it is still as breathtaking today as when it was completed nearly 200 years ago.*

## Principles of the Suspension Bridge

The basic principle of this type of bridge is that a deck is hung by vertical hangers from main suspension chains or cables, which fall into a catenary curve when supported by towers at each end. The cables are tied fast in anchorages secured in the ground beyond. The rods, cables, and chains are in tension as the deck pulls down on them, while the towers are in compression as they carry the dead weight and any live load.

The main advantage of this type of structure is that due to its relatively light weight it can cross a much wider gap than other forms of bridge. As it requires only two towers for the main span, there is less need for the expensive, dangerous, and time consuming process of building piers across the water. The main span is designed to cross the deepest part of a river or estuary so that the towers can be built on the shore or in shallow water, usually with just a cofferdam rather than a caisson to work in. The reduction in the amount of material needed and the construction costs also make suspension bridges more economical to build than other types.

The elements of a suspension bridge which determine how far it can span and its durability are the type of suspenders used, the fixing of the towers and anchorages, and the design of the deck. The development of new materials and structures to deal with these factors have led to bridges up to eight times longer than Telford's being built.

The first major examples used wrought-iron chains for the main suspenders, similar to a bicycle chain but with elongated links. These were arranged in sets of two or more along each side, with vertical wrought-iron rods connecting them to the deck below. Telford and Brunel pushed this material to its probable limits. It took the mass production of steel in the late 19th century and more recently the creation of special high tensile cables for the next major steps forward. Modern bridges use hundreds of fine cables massed together to form the main suspender, which is then encased in a tube for protection from the elements. These cables have a tensile strength ten times that of wrought-iron chains. As high tensile steel is brittle, the cables which make up the main suspenders are continuous, so that there are no joints which could make them crack.

Towers need to be set on a sound foundation, a task which is still difficult despite being built in shallow waters or even on land, as they take a huge load,

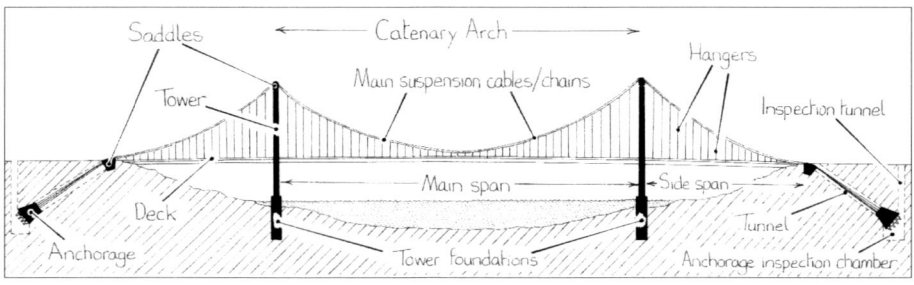

**FIG. 8.2:** *Diagram showing the main parts of a suspension bridge. The towers, which can be built in brick, stone, concrete, or steel, are in compression (black), while the cables and hangers, of either iron or steel, are in tension.*

# SUSPENSION BRIDGES

**FIG. 8.3:** *Examples of different types of main suspenders. The top photograph is of iron chain links used on the earliest major examples in the 19th century. These plates or links have pins through each end. The surrounding ring is of similar thickness to the main part of the bar, so as not to weaken it at the vulnerable connections. The second example shows a twisted cable used on some lightweight structures in the first half of the 20th century. The lower photo is of the outer protective casing of a large modern bridge, inside which are hundreds of individual cables running parallel with each other.*

and excessive movement in one can lead to the failure of the structure. Early examples tended to be massive stone and brick structures, whereas hollow steel or reinforced concrete towers are used today, though many have had additional strengthening inserted as traffic has increased.

The chains and cables can be fixed or pass over the tops of towers. Some form of saddle resting on a flexible bed is required to allow for slight movements due to expansion and the effects of wind upon the towers and suspenders. The ends of the chains or cables need to be fixed to a point either in tunnels dug down into bedrock or in a man-made chamber weighed down by a block of sufficient weight. (The search for suitable anchorage often means that the length and angle of the suspenders from each tower is different, and most suspension bridges are not symmetrical.) The fixing itself varies between each structure but in modern examples they are generally tied round the ends of rods which are set into stone or concrete inside a chamber, to permit checks for corrosion. In the tunnel a tapered shape is often used that is wider at the fixing and narrow nearer to the towers, so that a brick or metal plug built over the ends would prevent the suspenders being pulled out.

A problem for many suspension bridges is that either as part of the original design or as an attempt to improve fixings at a later date, the cables and chains have been set in concrete, making it difficult to assess the condition of the metal parts.

# Bridges Explained: Viaducts & Aqueducts

FIG. 8.4: *The first major bridges used massive stone or brick towers (left), while modern examples are taller, thinner hollow structures of steel or concrete.*

Despite being built only forty years ago the Forth Road Bridge is now facing the possibility of requiring new tunnels and fixings, as the original cables set in concrete cannot be accessed for inspection.

What has proved to be the most important aspect of the suspension bridge, that even the great engineers did not foresee, is the design of the deck, which has to be able to resist the live load and wind turbulence. If you ever get the chance to walk across a small footbridge of this type, try pacing down the middle and then return going along one side. You might be surprised by the increased movement this can create on the deck. Designers face a constant dilemma between making the deck as light as possible to reduce the strain on the cables or chains, and yet rigid enough to resist the movement created

by traffic and the weather. A heavy offset load like a train could start oscillations in the deck which in the worst situation could lead to complete failure. This happened on a number of early suspension railway bridges and the few which were built in this country tended to be quickly replaced by more rigid trussed structures. A famous recent example is the Millennium Bridge over the Thames in London, which began to sway dramatically, not because the designers had under-estimated the load but because people without being aware of it were walking in step, so that the rhythmic pattern of footsteps unbalanced the structure and strengthening was required.

Stresses can also be formed by cross-winds creating oscillations as a result of pressure applied to the face on the windward side and eddies around the other. Telford's Menai Bridge had its

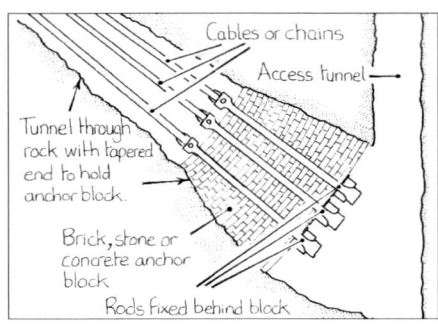

FIG. 8.5: *Diagram of an anchorage. Where there is good bedrock, it can be dug out, with a funnel shaped chamber, and the chains or cables set in a wedge-shaped block of brick or concrete permanently held back by the shape cut out of the rock.*

# SUSPENSION BRIDGES

deck wrecked in a storm only ten years after it was completed, and a new, stiffer deck had to be fitted. The most famous example of how even a relatively modern structure can fail was most dramatically caught by a passing cameraman one Sunday morning in Washington State, USA. The Tacoma Narrows Bridge, with a span of 2,800 ft, was built perilously narrow owing to an anticipated low volume of traffic. Almost as soon as it was opened, it was nicknamed 'Galloping Gertie', and it lasted only four months, as, on 7 November 1940, a moderate wind hitting the plate girder side of the deck caused oscillations which multiplied out of control until the deck ripped itself away from the cables and was sent crashing into the river below. The understanding of why this had happened, and the realization that probably it had happened before to numerous earlier structures, led to revisions of the hangers and to new types of decks being designed in the second half of the 20th century. Some increased the stiffening of the deck using open trusses rather than plate girders, which helped to dissipate the action of the wind; others looked to make the deck aero-dynamic with a profile which would be unaffected by the wind.

**FIG. 8.6: MILLENNIUM BRIDGE, LONDON:** *This modern suspension bridge was built by leading designers with the aid of computers and yet even they could not have foreseen how its popularity upon opening would cause it to sway alarmingly.*

## The Development of Suspension Bridges

The suspension bridge is known to have existed in Asia from ancient times, the Chinese even used metal chains for them by the 6th century AD. Simple timber plank decks precariously crossing gorges supported on bamboo rope or vines and swinging alarmingly are familiar from films, and were probably widespread in some parts of the world. In western Europe, however, there is nothing to suggest they were used. The first records appear in a design by Faustus Verantius dating from the early 17th century in which he illustrates a simple chain suspension bridge.

The first suspension bridge in this country was built in 1741, a 70 ft footbridge over the River Tees with wrought-iron chains but with the deck laid onto, rather than suspended from, them. (It was dismantled in 1802 but the chains are in the Bowes Museum at Barnard Castle.) It was not until the early 19th century that the modern type of bridge took shape, principally in America with the designs of James Finley who, long before most, realised that the deck had to be level and braced with trusses to keep it rigid. It was here that the first bridge using cables rather than chains was built in 1816, and were further developed in France and Switzerland (wire was cheaper in these countries and had greater tensile strength than chains). That was until 1850 when soldiers crossing the Basse-Chaine Bridge created oscillations in the deck which caused it to collapse thus undermining confidence in this type of structure. It was back in America where wrought-iron and later steel wires spun into cables supported spans over 1,000 ft (nearly 1,600 ft on the Brooklyn Bridge opened in 1883).

In this country we preferred to use wrought-iron chains, which were considered less susceptible to corrosion and could be erected in shorter sections. It was the work of Captain Brown who patented a new type of link in 1817 which was crucial. He designed the famous Brighton Chain Pier and in 1820 built the Union Bridge across the River Tweed, with a span of 360 ft linking England and Scotland. It was Brown who Telford consulted when he first began to design his suspension bridge over the Menai Straits.

## The Menai Suspension Bridge

In the wake of the Irish Rebellion of 1798, in which around 20,000 lost their lives, the British Government sought control and passed the Act of Union, permitting selected Irish MPs to sit in the House of Lords. To further reinforce tighter links and more effective control over Ireland, a body was formed to improve communication by building a new road from London to Holyhead, the main port to the province. With the end of the Napoleonic Wars in 1815, the commission set to work, appointing Thomas Telford chief engineer. For most of its route through England short bypasses, new bridges, and improved surfacing were all that were required but, for Wales, a far more ambitious scheme was planned, with a new road cut through rock and clinging to the side of gorges as it wound its way to Bangor on the coast without a gradient steeper than 1 in 22! This incredible piece of engineering has been rather lost to history because as soon as it was completed the railways stole the limelight and the line to Holyhead built by Robert Stephenson only 20 years later took most of its traffic.

Nevertheless, this did not diminish the highlight of the route, the crossing of the Menai Straits to Anglesey. A bridge here had long been considered as an alternative to the unpopular and dangerous ferry (thousands of cattle had to swim the straits each year to reach market), and the great bridge-builder John Rennie (see Chapter 11) had already proposed a cast-iron structure in 1801. Telford initially came to the same conclusion, but that was without the consent of the Admiralty. Their ships used this tidal passageway and they insisted that any bridge would have to permit the passage of ships that were not only tall but also wide. With arched and beam types dismissed (there was no central island to rest a pier), this left Telford with no choice but to build a suspension bridge. Such a structure had been built before, but not on this scale.

Construction of Telford's masterpiece began in 1818 with preparation of the site, the building of quays, laying of railways, and erection of workshops. Ynys Y Moch, the small rocky outcrop upon which the Anglesey tower would be built, was also levelled at this time. Tunnels were then dug for the chains to run down, with horizontal chambers opened up at the bottom, in which the cast-iron frame that would hold them fast was built. On the mainland the chains vanished into a large stone building; from here they went down into the ground, where they were embedded in large masonry blocks. On the Anglesey side the chains had to be angled over a saddle in order to reach down to a suitable fixing. A door on the south embankment led to this deeper chamber. The approaches were built as conventional masonry viaducts. The towers were built with a hollow upper section, that was internally braced rather than filled with rubble in order to keep the weight down. They extended above the deck level, with twin portals for traffic to pass through. Metal saddles on the top carried the chains and allowed for slight movement.

FIG. 8.7: MENAI SUSPENSION BRIDGE, GWYNEDD: *Telford's bridge today has four steel chains replacing the original sixteen wrought-iron sets.*

FIG. 8.8: MENAI SUSPENSION BRIDGE, GWYNEDD: *The building on the left stands at the head of the mainland archorage tunnel, while the deeper Anglesey fixing has to be accessed via this doorway (right) in the embankment.*

turned until the chain had reached the top of the tower, a process which took under two hours. The other fifteen chains (in all there were four sets of four) were fixed in a similar way. Suspender rods and the metal and timber deck were then added in time for the first crossing, by a mail coach in January 1826.

For all his planning, Telford, like many other great engineers, misjudged the effects of the wind upon the deck. There were no wind tunnels in his day, and a true understanding of aerodynamics only came when man took to the air some 70 years later. The deck was damaged badly by a storm in 1839; it was strengthened and yet needed

The viaducts and towers were complete by 1825, with the side-span chains fixed in place. The chains had been cast in Ironbridge and tested at Shrewsbury for loads in excess of what was anticipated, before they were transported to Chester and shipped round the coast to the site. Telford was meticulous in his preparation, but the part which gave him most concern was the lifting into place of the 23-ton chains of the main span, which had to be done quickly because of the tides and so as to avoid blocking the navigation any longer than necessary. One half of the main span chain was fixed to one of the towers and left hanging down, while a wooden platform with the other was moved into place between the towers. The two lengths were pinned together and the unconnected end was tied to ropes fixed to large capstans, which were

FIG. 8.9: CONWAY SUSPENSION BRIDGE, GWYNEDD: *Telford designed this 325 ft span bridge to take on the appearance of a drawbridge to the adjacent castle.*

# Suspension Bridges

**FIG. 8.10: CONWAY SUSPENSION BRIDGE, GWYNEDD:** *Detail of the iron chains passing over the saddle and into the ground. Concrete poured into these tunnels in the 1890s is now of concern, as the condition of the chains within is unknown.*

repair again before the century's close. The 16 original iron chains were completely replaced by four steel chains when the bridge was strengthened in the late 1930s, prior to tolls being abolished. Today, despite the changes to its original form, it is still a magnificent structure, awe-inspiring from the water's edge and hair-raising to cross!

## Conway Suspension Bridge

At the same time as the Menai Bridge was being constructed, a smaller scale structure, also designed by Telford, was taking shape directly below the castle at Conway. The towers supporting the iron chains were built to mirror their medieval neighbour, cylindrical in form with battlements. In this case the chains on the castle side run straight into the rock face beneath it, making an easy fixing point. On the other side they pass down through tunnels to reach bedrock. The rods hanging down from the chains to the deck are made in two lengths with a screw-threaded turnbuckle fitted halfway down to link them together and allow adjustments to be made to the tension.

**FIG. 8.11: MARLOW BRIDGE, BUCKS:** *Another engineer who successfully constructed suspension bridges was William Tierney Clarke. His Marlow Bridge of 1829 still carries traffic over its 217 ft span across the Thames. It is supported on iron chains hung from classically styled towers. He went on to greater fame with a bridge spanning over 600 ft linking Buda and Pest in Hungary, after which the city became known as Budapest.*

# Bridges Explained: Viaducts & Aqueducts

This bridge is notable because, unlike the Menai Bridge, it still has much of its original ironwork, thanks in part to its replacement in 1958 by a steel arched bridge. When it was restored in the 1990s by the National Trust, nearly 150 tons of hardcore which had been built up over the years was removed from the deck. This relief in stress on the chains was fortunate, as the metal in some links was down to a twentieth of its original diameter. The timber deck of the bridge was removed some time ago but pieces were cut up to make a new floor in the adjacent toll house, which can still be viewed today.

**FIG. 8.12: CLIFTON SUSPENSION BRIDGE, AVON:** *This iconic bridge was originally designed with exotic Egyptian style towers, but the project was plagued with financial problems and they were finally completed without the full decoration.*

## Clifton Suspension Bridge

When his Menai Bridge was completed in 1826, Telford had felt that the span of 580 ft was about as great as was possible with wrought-iron chains. Only a few years later a young Isambard Kingdom Brunel was to prove him wrong, with his plan for a bridge of over 700 ft across the Avon gorge at Clifton.

The turbulent story of Clifton Suspension Bridge spans over 100 years. It was as far back as 1754 that a Bristol merchant, William Vick, left £1,000 in his will to the Merchant Adventurers for the building of a stone bridge across the gorge. By the 1820s the boom in the city from trade, together with the £8,000 which Vick's bequest had grown to, led to an Act of Parliament being passed to build a suspension bridge. (A stone one was estimated to cost around £80,000.) A competition was held in 1829, with an aging Telford as a rather biased judge who ended up rejecting the other entrants in preference to his own design. His, however, being too expensive resulted in a further competition the following year, in which Brunel's plan was chosen. Unlike Telford's design, which had huge Gothic towers rising from the river valley to reduce the length of the main span, Brunel's bridge placed towers up the sides of the gorge, making it much cheaper to construct but requiring a greater span.

Problems began almost straightaway as rioting in Bristol in 1831 caused commercial turmoil, and work did not get fully underway until 1836. Then, when the towers were almost complete,

# Suspension Bridges

the money ran out and the project went on hold. Brunel carried on with other work, including the Hungerford suspension footbridge across the Thames near Charing Cross and the Royal Albert Bridge at Saltash, which, ironically, used ironwork that had been destined for the Clifton Bridge.

When Brunel died in 1859 all that stood were the towers on each side of the gorge, but a sympathetic public and grieving engineering colleagues spearheaded a new drive to raise funds to complete the project as a memorial to the great man. In another twist Brunel's Hungerford Bridge was being demolished to make way for a new railway bridge (using the original piers, which still support this bridge today), so that the chains from this structure were purchased in order to complete Clifton. With slight revisions the bridge was opened in 1864.

The towers were built in the fashionable Egyptian style, which Brunel favoured, and the southern one

**FIG. 8.13:** *A close up view of the iron chains. The links are offset and individual rods from each one form a tightly packed series of hangers, giving the deck more rigidity. The wider sections on the hangers are turnbuckles, which are rotated to adjust tension.*

**FIG. 8.14: CLIFTON SUSPENSION BRIDGE, AVON:** *View of the bridge from the river, showing how the towers were built up on stone bases high up in the wall of the gorge.*

# BRIDGES EXPLAINED: VIADUCTS & AQUEDUCTS

**FIG. 8.15:** *Examples of metal suspension footbridges dating from the first half of the 20th century. Top left is Queen's Park Bridge, Chester, which was opened in 1923, replacing a suspension bridge built some 70 years earlier. Top right is a smaller structure from Betws-y-Coed, and bottom left a similar bridge over the Dove near Uttoxeter. Bottom right is Porthill Bridge, Shrewsbury. Although built in 1922, it originally had chains, now replaced with steel cables. Walking across bridges like these can create many of the movements which affect larger examples.*

is built on a huge stone abutment, which was originally believed to be solid but only as recently as 2002 was found to contain hollow chambers linked by narrow tunnels! The three chains on each side are offset so that the hangers which fall from each link are close together, helping to make the deck more rigid (Fig 8.13). The shape of the link is important, with the depth of material around the bolt the same as that of the bar, so that tension is equally passed along it. The curved ends avoid concentrations of stress, which occur in right-angled corners (Stephenson's Dee Bridge probably collapsed in 1848 because of a sharp corner in a decorative part of the girder). It is important to keep all three chains in perfect tension – if one is only slightly shorter, it may take all the weight – and so they are independently hung, with an

# Suspension Bridges

adjustable turnbuckle on each hanger to correct the tension. The frame of the deck is made of wrought-iron girders, originally with an underside of 5 ins thick timber and another of 2 ins thick timber transversely across it.

In the second half of the 19th century few new roads were planned in view of competition from the railways, and suspension bridges, which were unsuitable for this latter form of transport, were rarely built. Exceptions are to be found mainly in London, where traffic increased, and pressure to cross the Thames resulted in the Hammersmith and Albert bridges. More common were lightweight footbridges with iron or steel towers and cables, which became especially popular in the early 20th century.

## Forth Road Bridge

It was not until after the Second World War, when the focus had swung back to the road network, that large rivers and estuaries had to be crossed again. This time, unlike the railways, which had used metal trusses on numerous piers to create a rigid, strong bridge, cars and lorries with a lighter and widely displaced load could be carried on more economical suspension bridges. The first of these was planned to cross the Firth of Forth a short way upstream from the famous cantilevered rail bridge (Chapter 7).

Since the great chain suspension bridges of the 19th century, developments in America and around the world had changed the way in which they were constructed. In this case, with the huge 3,300 ft main span, the towers could be set up in the shallow water at the edge of the deep channel, with steel sheeting used to make a cofferdam and piles driven to the bedrock some 100 ft below. Rather than masonry or cast iron, the towers of the Forth Road Bridge were built in steel box-sections with diagonal cross-bracing, one fitting on top of the other until the height of over 500 ft was reached. These hollow structures were lighter but stronger and allowed access to the top for maintenance.

FIG. 8.16: FORTH ROAD BRIDGE, LOTHIAN: *It was engineering companies rather than great individuals which were responsible for major 20th-century bridges. One of the most renowned is Mott, Hay, and Anderson, who, along with Sir Freeman Fox and Partners, were responsible for the Forth Road Bridge.*

# Bridges Explained: Viaducts & Aqueducts

The huge anchorage chambers were tunnelled out of the rock at each end, with a mass of concrete used to hold beams in place onto which the yokes were bolted. The cables were made up of more than 11,000 individual high-tensile steel wires formed into large strands, which splayed out at the end so that they could be fixed onto the yokes. They were hung across the towers by wheels spinning on a continuous loop from one end to the other, a system which was capable of laying out more than 500 miles each day, the total in both eventually reaching 31,000. The completed main cables had suspenders hung from them and were then covered with an outer casing for protection.

The deck was the first major structure built here following the Tacoma Narrows Bridge disaster in the United States. It had to be light, but strong enough to be rigid, and open enough to resist the wind pressure in this exposed position. The designers used large steel trusses with hollow box-sections top and bottom, and ladder-type girders between, to keep the weight down and yet be open enough to allow the wind to pass through without causing the oscillations which ripped the American bridge to pieces. Gaps between the carriageways on the upper surface also aided this.

Although delayed by problems with the weather, the bridge took six years to complete, opening in 1964. Its total length is 8,259 ft, including the concrete viaducts at each end. The dramatic increase in traffic beyond that projected at the time has meant further strengthening and repairs have been needed in recent years. Columns have been inserted inside the towers and hangers have been replaced. Excessive movement in the towers was experienced and countered using weights hung from the main cables as a dampening system. There are now problems with corrosion and, as the rods in their concrete anchorages cannot be inspected, new ones may

**FIG. 8.17: TAMAR ROAD BRIDGE, CORNWALL:** *Built from 1959 to 1961, this similar structure to the Forth Road Bridge, with an open truss supporting a reinforced concrete deck, had a main span of 1,100ft and was the longest bridge in the UK at the time. It was built for the local public authorities who ran out of patience with the government and formed a committee to oversee construction once the necessary Act of Parliament had been granted in 1957. It is more notable now for the ingenious way in which it was recently widened by cantilevering an extra lane on either side without greatly increasing the load or closing the bridge to traffic.*

# Suspension Bridges

have to be built at a huge cost. At the same time, the long traffic delays have made a completely new bridge a serious consideration, although ironically a hovercraft ferry service has now been established to help take the pressure off the current structure.

## First Severn Crossing

Initial designs for a suspension bridge across the Severn Estuary were produced in 1946 by the team that was responsible for the Forth Road Bridge, although this latter project took priority and was built in a similar style to their planned Severn Bridge. It was Sir Gilbert Roberts of Freeman, Fox, and Partners who was responsible for the new design when their attention returned to the Severn Crossing.

The Forth and Tamar bridges had followed the American lead in using large open steel trusses to stiffen the deck against wind pressure, but Roberts looked to Europe for inspiration, where Fritz Leonhardt had suggested making the deck of a aerodynamic profile so as not to disrupt the wind flow. After tests with various designs, a distinctive thin angular shape was chosen which would not only resist the eddies that applied torsion to the structure but also direct the wind to apply a downward force on the deck (which the cables could hold in check) rather than an upward one (which they could not). This revolutionary deck design was formed out of box sections which were prefabricated in nearby steel works and shipped to site, where they were lifted off the barge and fixed in place.

**FIG. 8.18: THE FIRST SEVERN SUSPENSION BRIDGE, AUST, SOUTH GLOS:** *Opened in 1966 after taking five years to build — the building of the piers was hampered by the strong Severn tides — it had a main span of 3,240 ft, only just short of that of the Forth Bridge, but thanks to its revolutionary design it is a great deal lighter and required far less steel.*

There were other differences from the earlier project, most notably in the arrangement of the hangers. These were not vertical as at the Forth Bridge and its predecessors but hung in a zigzag diagonal arrangement, designed to increase the rigidity of the deck by their triangular form (Fig 8.19). The towers, like the deck, were lighter and were made from steel plates formed into two tall box-posts with internal bracing and

# Bridges Explained: Viaducts & Aqueducts

FIG. 8.19: *A view from the deck showing the slope between the footpath and the road which is taller on the top than a similar one below. When the wind strikes it from the side, it creates a greater force on the taller top slope which pushes the deck down slightly, helping to stabilize it.*

FIG. 8.20: THE FIRST SEVERN SUSPENSION BRIDGE: *Owing to the unsuitability of the adjacent rock, the anchorages were built in huge concrete chambers, over which the deck runs at each end. The eastern anchorage (centre bottom in the above photo) is on the river's edge; the western is on dry land.*

three horizontal cross-members. As in the case of the Forth Bridge, however, increased traffic has led to the insertion of steel columns to relieve some of the load. These design factors give the Severn Suspension Bridge a graceful elegance missing from some of the trussed deck versions.

## Humber Bridge

This largest of British suspension bridges followed the lead of the Severn Bridge in using an aerofoil deck and diagonally arranged hangers. It also had huge concrete anchorage chambers built on the shore to hold the cables in a

FIG. 8.21: HUMBER BRIDGE, HUMBERSIDE: *The huge anchorage on the Barton side weighs up to 300,000 tons. Inside the hollow structure the cables are fixed to steel rods which pass through the concrete anchor blocks.*

# Suspension Bridges

**FIG. 8.22: HUMBER BRIDGE, HUMBERSIDE:** *This is the largest single-span bridge in Britain; from the time it was opened in 1981 until 1998 it was also the longest suspension bridge in the world. Its total length between the anchorages is 7,300 ft, with a main span of over 4,600 ft. The side spans are uneven, the 1,750 ft southern being nearly twice as long as the northern. Because of its low clearance — it was expected that there would not be large ships using the waterway — the towers are proportionally shorter than in most similar structures, making the bridge appear even longer.*

similar manner to the fixings on the Forth and Severn bridges. But unlike these the Humber Bridge had reinforced concrete rather than steel towers, each built on twin caissons.

Despite its huge 4,624 ft main span, the southern tower had to be built in the water and this was the main cause of delay in building the bridge, although political wrangling had already set the project back. It had originally been planned in the 1930s, and an Act for a suspension bridge was passed in 1959, yet the final go ahead was only given in 1969, with building further delayed until 1972. When the actual spot for the southern tower was investigated, it was found to be on shifting sands, and the best site which could be found offered only a hard clay bed and so the foundations here are four times deeper than the 25ft of the north tower. The southern anchorage is also heavier, as it is dug into the clay (unlike the northern one, which is set in a hard bed of chalk) and weighs an immense 300,000 tonnes. The bridge was finally opened in 1981 and, 20 years after its completion, it had already had more than 100 million vehicles pass over it. There has long been a tradition of carving good luck symbols or burying something (or even someone) in the foundations of bridges. The Humber Bridge continued the tradition, not with a sacrificial offering

# Bridges Explained: Viaducts & Aqueducts

**FIG. 8.23: HUMBER BRIDGE, HUMBERSIDE:** *The underside of the deck showing its aero-dynamic profile and the cantilevered cycleway on each side.*

but with a time-capsule set in the base of one of the caissons.

The Humber Bridge is the last major bridge of its type built in Britain so far, although more are planned. In the past three decades a new form of bridge has been developed for the latest major river crossings. Suspension bridges were previously chosen for their economy, as they avoided the need for establishing numerous piers in turbulent water or on a soft seabed. Now, with modern machines, materials, and technology, building long viaducts over wide stretches of water using hollow concrete beams resting on piers is no longer such a dangerous and expensive problem. However, there is still a limit to how far a concrete span will stretch and, as most large waterways need to permit shipping as well as having a deeper

channel in the middle, another form of bridge is still required for the central section. The type now chosen in this situation is often the cable stay.

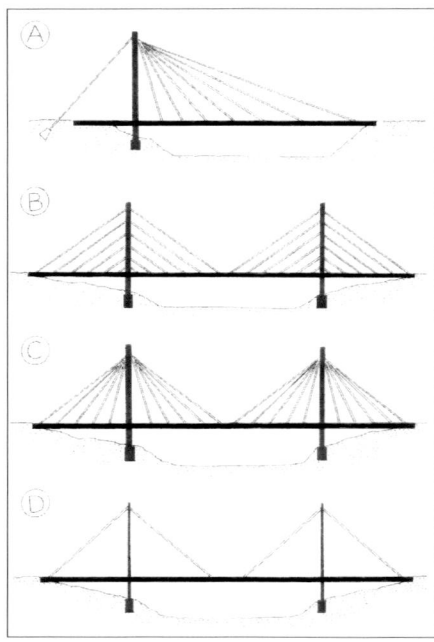

**FIG. 8.24:** *Diagrams of different forms of cable stay bridges. Some have an offset tower with the cables anchored to one side (A), others are balanced, with the deck in effect cantilevered out from each side (B-D). The cables tend to be arranged in a harp pattern (B) or fan design (C), although some are in a hybrid arrangement (Fig. 8.29). As in a suspension bridge, the cables are in tension and the tower (outlined black) is in compression, but in a cable stay bridge the deck also needs to resist compressive forces, as the cables are pulling it towards the tower.*

## CABLE STAY BRIDGES

A stay is a support which keeps a member vertical, like the diagonal ropes used to hold a ship's mast or a flagpole upright. In the cable stay bridge the role is reversed, as here the vertical member is a tower or pylon from which the stays drop down at an angle to hold up the deck.

The cables can be arranged in many ways, although there are two principal methods. A harp design is where the cables run roughly parallel to each other from a point sequentially higher on the tower as they reach further out along the deck (Fig 8.24B). A fan pattern is formed when the cables are fixed to a single point on the tower and radiate downwards to various points along the deck (Fig 8.24C). Earlier types, from the 1960s and 70s, tend to use fewer, thicker cables or even a single cable (Fig 8.24D), but problems such as sagging in the unsupported length of the deck and the extra cost of this type means that most recent examples use multiple thinner cables. For this type of bridge the cables can be manufactured away from site and brought in when needed, unlike suspension bridges, where the cable is spun on site, a much more expensive and time-consuming process.

Although it is referred to as a suspension type, the cable stay bridge is different in a number of ways, making it a more economical choice. The suspension bridge is held between two towers, which vary in height depending upon the span and clearance above water of the deck. The cable stay bridge is more flexible and can be supported by one, two, or a series of towers (sometimes called pylons). A single support is often used for smaller bridges with a light load, such as a footbridge or minor road bridge, and has the advantage of being cheaper than a conventional two-tower suspension bridge; in addition, its support can be sited on the best foundation, whichever bank this is on.

Larger structures, like those which carry motorways over wide rivers, usually have a tower at each end. These can either be a single column with the carriageway held each side (Fig 8.27) or a double with the road running between (Fig 8.29). An advantage of these larger cable-stayed bridges is that the deck is balanced either side of the towers during construction, meaning that the roadway can be built out in sections from the centre point evenly on both sides (see Fig 8.24 B-D). This is a much quicker and cheaper method than

FIG. 8.25: JACKFIELD BRIDGE, IRONBRIDGE GORGE, SALOP: *A modern cable stay bridge with a single tower; the cables anchored in the southern abutment, behind it.*

having to spin extensive cabling back and forward before deck construction can begin on a conventional suspension bridge. Another big advantage is that no anchorage is required in this balanced type, something which is often an expensive problem, as in the case of the Humber Bridge. (Single tower bridges can be balanced, but usually have an anchorage point on the landward side.) The pier at the end of suspension bridges is often used to redirect the main cable towards the anchorage, which applies a load meaning that the pier has to be substantial. On a balanced cable stay bridge, this pier supports the deck from the minimal extra weight of the passing traffic only, and hence is smaller and therefore cheaper to build.

The deck of a cable stay bridge, unlike that of a suspension bridge, is under compression from the diagonal cables, which tend to pull it back towards the tower. This is a problem during construction until the cantilevered ends are finally fixed together. To counter it the deck needs to be stronger to resist the buckling effect, although this, together with the shorter spans and the way the deck is fixed, makes the cable stay bridge more rigid than conventional suspension types and less prone to the stresses caused by wind and the live load.

## The Development of Cable Stay Bridges

As is usually the case with bridges, this type is not a new invention; designs were drawn up, but not executed, as far back as the early 19th century. Stays were used in a number of suspension bridges, most notably by John Roebling on his Niagara and more famous Brooklyn Bridge, to provide additional strengthening to keep the deck rigid. The Albert Bridge, across the Thames at Chelsea, uses both a main suspension cable with hangers and separate iron chain stays to support its roadway.

It was not until after the Second World War, with the development of high tensile cable and reinforced concrete, that a bridge supported by stays alone was considered. A number were built on the Continent, including a set of three over the Rhine at Dusseldorf, designed by Fritz Leonhardt. The first major example in this country was at the western end of

**FIG. 8.26: ALBERT BRIDGE, LONDON:** *This ornate bridge with its column-encrusted towers and decorative ironwork so beloved of the Victorians was designed by R.M.Ornish and completed in 1873. Its 400 ft span across the Thames is supported by a fan arrangement of iron chain stays, the main catenary arched cable and its hanging rods taking a secondary role.*

the Severn Suspension Bridge, which after crossing the main river has then to span the smaller River Wye. This was built with a pair of single towers in the centre of the carriageway, which was supported by a large single cable. With so few supports, sagging of the deck occurred in the large gaps, and the bridge had to be modified in the late 1980s with a twin set of strands and a more open arrangement, so that they could be easier to inspect (Fig 8.27).

Cable staying has come into its own in the last few decades, not only in its widespread use by engineers for large-scale projects but also in the hands of architects, who, like their predecessors in the late 18th century, have turned their attention to the decorative possibilities of bridge design. This is most commonly seen in the wide range of footbridges to be found in all parts of the country: cheap-to-build steel structures with a dash of flair. They use the great strength of steel to make slanting towers, curved decks, and sweeping arrangements of cables, which seem to put the bridge under unnecessary strain for purely decorative reasons. Yet, with computer-aided design, they keep within the boundaries of stresses and create iconic symbols which are a bright and entertaining break from the somewhat monotonous concrete structures.

It is, however, for large-scale projects that cable-staying is now becoming famous around the world, especially as the arrangement of cables and the simple form make these structures most attractive. This form of bridge cannot cover as wide a span as a conventional suspension bridge and so, over deep or large expanses of water, the latter is still used. Concrete box-girder bridges are still the most economic for narrower or shallower crossings. It is for structures between these sizes that the cable stay has become dominant, although it is usually as part of a much longer concrete viaduct, like the incredible Oresund Bridge, which links Denmark to Sweden and is four and a half miles long, with a cable-stayed central span of 1,600 ft. In this country the Second Severn Crossing is more than three miles long, with a central navigation span over the deepest channel of 1,500 ft.

The spans of stand-alone cable-stay bridges are getting longer. The Pont De

FIG. 8.27: WYE BRIDGE, GWENT: *The original single cables were replaced with this new arrangement of cabling consisting of two sets of twelve, held in an open arrangement for ease of access for inspection.*

# Bridges Explained: Viaducts & Aqueducts

**FIG. 8.28:** *Examples of cable stay footbridges with single towers, (clockwise from top left) at Lancaster, Shrewsbury, Worcester, and Manchester.*

Normandie, near Le Havre, completed in 1994, has a span of 2,808 ft, while the Stonecutters Bridge, Hong Kong, which is planned to open in 2008, will span a 3,300 ft gap. These still fall short of the possibilities of conventional suspension bridges; one proposed crossing of the Messina Strait between Southern Italy and Sicily is planned to have a main span of 10,000 ft. Of all the recent cable-stay bridges, the one which seems to have caught the imagination is one which is seemingly the least necessary, bridging a valley which surely could have been crossed by more conventional means. Yet this great pioneering spectacle, the Millau Viaduct in France, defies belief, with its elegant cloud-piercing pylons as tall as the Eiffel Tower carrying the A75 at nearly 900 ft. It was designed in part by British architects Fosters and Partners.

# SUSPENSION BRIDGES

## The Queen Elizabeth II Bridge

The first of the modern large-scale cable-stayed bridges in this country was built to carry the M25 across the Thames near Dartford. (The Erskine Bridge, near Glasgow was the previous largest, completed in 1971, with a 1,000 ft main span, and a similar form to the Wye Bridge.) The existing tunnels were understandably unable to cope with the dramatic increase in traffic generated by London's new orbital motorway, and so legislation was put in place and construction of a new bridge began in 1988. The Thames at this point is still a busy waterway, and a clearance of nearly 200 ft in height and virtually the full width of the river was required so that even large liners could pass beneath. At the same time it had to support one of the busiest roads in the country, making a cable-stayed bridge the best choice. (It takes only the four lanes of southbound traffic, the northbound using the existing tunnels.)

As the towers were being founded in the shallower waters near the edge, a pair of huge concrete caissons, which could be simply dropped in position to form in effect a cofferdam, were built in Holland and shipped across the North Sea. They could be filled in with concrete to form the base of the towers on site, creating a mass of some 85,000 tonnes each. On top of these the main concrete piers were built to a height of 175 ft. Erected on these were thinner steel towers of a further 275 ft, which supported the cables in a hybrid fan-harp arrangement. The main span of

**FIG. 8.29: QUEEN ELIZABETH II BRIDGE, DARTFORD, KENT:** *This huge cable stay bridge completed in 1991 had to give a clearance to shipping of nearly 200 ft, making for steep inclines on each side.*

## Bridges Explained: Viaducts & Aqueducts

nearly 1,500 ft is not symmetrical between the towers. Owing to the positioning further into the river of the north one, the span peaks slightly closer to this tower to keep the maximum clearance in the centre. Despite the large scale of the operation, it was completed on time and opened in 1991.

### The Second Severn Crossing

Around the same time as this project was coming to a close, another motorway crossing – this time on a much larger scale – was being planned across the Severn Estuary. Pressure from traffic using the existing Severn Suspension Bridge was such that another crossing was needed: one which would be stiffer, so as not to limit use so much when the wind speed was high. The site chosen was a few miles further south of the earlier bridge and, although the width of the river here was substantially greater, it could make use of outcrops of rock on both sides of the estuary.

Building the long viaducts up to the main span crossing the deep navigation channel was made possible only by using GPS systems and large modern lifting-vessels which were capable of moving and positioning the huge single piece caissons onto site in the limited time determined by the strong tidal range. The concrete lozenge-shaped caissons were prefabricated at the river bank works. They were greater in height than the high-water level so that they would act as a cofferdam when positioned. They were carried out to site at high tide on huge barges powered by independent underwater thrusters, which could be controlled by GPS to position the load within a couple of feet of the drop zone. A massive crane fixed to an oil rig type structure would then lift each caisson up slightly without moving its horizontal position, so that the barges could sail back. When the tide had gone out the caissons could be lowered onto the prepared river bed, where they were then sealed and the work of filling with concrete could begin, unaffected by the tides.

The piers were built up in sections on the completed caissons, with a vertical duct running through them containing stressing cables, which when tightened provided additional strength. The 200 tonne concrete segments which make up the individual carriageways were built out as a balanced cantilever from

**FIG. 8.30: THE SECOND SEVERN CROSSING, AVON:** *This immense bridge is nearly 17,000 ft long, although the cable stay main bridge is 3,100 ft.*

# SUSPENSION BRIDGES

each pier, with a final link between the two, in effect the central reservation, fitted on site as they went along.

In the centre was the main cable-stayed span, supported from concrete towers or pylons, with two crossbeams to hold the tapering columns. The deck of this part was made from prefabricated sections of steel with a concrete roadway and was built out symmetrically from each side of the tower. The sections were lifted off the barges by cranes fitted to the end of the previous section. Each section was held in position while it was bolted and sealed and the cables connected, before the cranes were moved to the section end for the process to be repeated. Despite the limited periods in which construction could take place owing to the tides, the whole three and a half mile structure took only four years to build, opening in 1996.

This permanent crossing is the latest method of bridge-building across large rivers. It succeeds arches, trusses, and suspension bridges as the most suitable method to give clearance for shipping and to create ever longer spans. There is, however, a problem with all these types and that is that, in order to create the necessary clearance for shipping, they need to gain height on land before spanning the water. In certain situations, especially in built-up city centres, long abutments to make this possible are not practical or the land required is simply not available. To span large rivers in these locations it would take something more flexible, to be exact, something moveable, literally a moving bridge.

**FIG. 8.31: THE SECOND SEVERN CROSSING, SOUTH GLOS:** *The central bridge is a cable stay structure with a main span of 1,500 ft.*

# Chapter 9

# Moving Bridges

**FIG. 9.1: TOWER BRIDGE, LONDON:** *Stretching nearly half a mile, including its approaches, with a main section of 880 ft and central span of 200 ft, it was the largest lift bridge in the world (it is actually a composite structure using suspension and cantilever principles) when it was opened on 30th June 1894.*

## Lift Bridges

Lift bridges were useful in flat areas where a waterway with traffic had to be crossed and where, whether to keep costs down or because there was no room for tall abutments, a conventional fixed structure was not practical. They consist of a framework supporting a mechanism which is used to raise a single or double leaf, with the deck either pivoted or lifted horizontally.

The important part in the design of any lift bridge is how to raise the deck, which, even in a simple structure, must be strong enough to take the weight of the live load. If it was simply hinged,

# Moving Bridges

the deck would be too heavy to be moved easily. The problem is solved by using a counterweight – a force a fraction less than that of the deck – being applied at the opposite end of the hinging point, so that a minimal pressure can make it lift or swing. Sometimes this crucial element is obvious other times it is not.

The earliest significant use of a lift bridge in this country was the castle drawbridge (also called a pont levis or turning bridge). It was not until the late 13th century, almost as the castle began to decline, that the drawbridge became popular. Previously, moats were usually bridged with fixed timber or stone structures, some with an area of planks which could be removed to create a gap at time of trouble. The earliest drawbridges were supported from above by chains hanging from the ends of horizontal beams which were hinged at the wall, with counterweights at the opposite end. The later, more familiar type was hinged at the castle side while the far end was raised by chains wound up by means of a capstan or windlass in the tower above (known as a bascule from the French for seesaw). Although no original examples survive, examples of the chamber in the base of the tower where the counterweights would have been can still be seen.

Lift bridges are likely to have been used to cross various small waterways in the past but it was not until the 18th century, with the increase in constructing river navigations, docks, and, most importantly, canals, that they became common. They were ideal as a cheap alternative to the brick arch

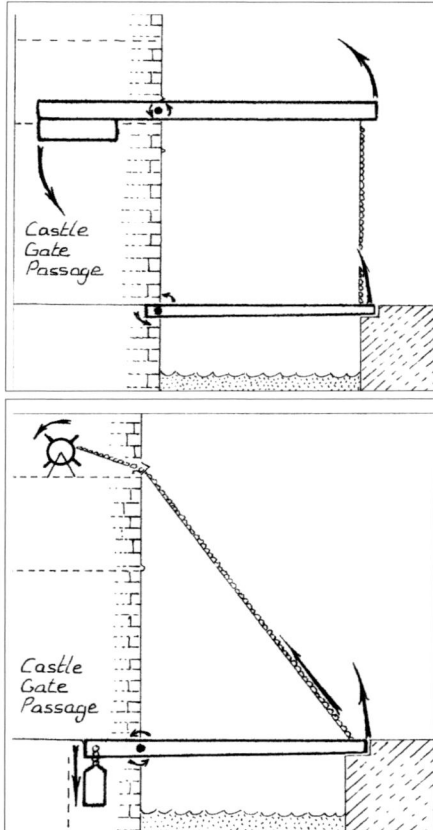

**FIG. 9.2:** *Diagrams of two types of castle drawbridge, with counterweights on the short end of the pivoted beams.*

bridge where its strength was not necessary (for footpaths, farm tracks, or minor roads crossing these narrow waterways, for example), or when the abutments required could not be fitted in. They were also useful as they could be left open or closed depending on which route was the busier.

Most had the hinged deck supported at the opposite end by chains hung from a balanced beam above, so that a slight pull on a chain or capstan at the far end would raise the bridge. In some situations the deck could be raised horizontally from the centre or from each corner, on towers with cables or chains attached to counterweights. This is a form which has been used for road bridges in recent years from Kingsferry, connecting the Isle of Sheppey to the mainland, and at Salford, across the Manchester Ship Canal.

## Tower Bridge

Although it does not appear so today, the Pool of London from the Tower upstream to London Bridge was a busy centre of shipping in the mid 19th century. The booming metropolis, however, was expanding to the east and south, increasing the pressure for another bridge further downstream from the ancient crossing. (At a census

FIG. 9.3: WRENBURY, CHESHIRE: *A lift bridge over the Llangollen Canal, shown with the deck down and raised.*

FIG. 9.4: LOCOMOTIVE BRIDGE, HUDDERSFIELD, YORKS: *This bridge from 1865 is connected in the centre by chains hung over the large wheels to a counterweighted rod on the left side, which when rotated raises the deck horizontally, supported by the black frame.*

# Moving Bridges

in 1882 around 22,000 vehicles passed over London Bridge in one day alone.) Despite opposition from many parties, especially the Lord Mayor of London, the authorities eventually had to concede and invited engineers and architects to put forward designs for a new bridge. It would need a clearance of 140 ft for tall ships to pass beneath and provide a route for a continuous flow of traffic above (a tunnel had already been opened below the site in 1871, but restricted access meant it had little effect). Because of the limited space on either bank, it could have only a short approach and not too steep an incline, so heavy carts could cross. It was also a requirement that the style of the adjacent Tower of London be respected, so that the bridge would complement the ancient landmark.

Various schemes were put forward, most being either bridges with movable sections, which generally failed by not providing a continuous flow of traffic, or high arched or truss bridges, which did not have sufficient clearance, had too steep an approach, or restrictive elevators to reach the actual crossing.

The leading contenders were the engineer Sir Joseph Bazalgette, with a number of fixed span bridges, including an elegant arch suspended roadway, and the architect Sir Horace Jones, with a medieval-style drawbridge and steel arch. The latter scheme seemed to be the only one which fitted the demands of all parties, although Jones had enlisted the engineer John Wolfe Barry, son of the famous architect Charles Barry, who designed the Houses of Parliament, to redesign the scheme before it was accepted. In this final plan the drawbridges pulled up by chains were replaced by pivoted bascules and the metal arch was changed to high-level walkways, which permitted pedestrians to cross while the bridge was raised. Unfortunately for Jones, just as he reached this potential peak in his career, he died suddenly in 1887 before the bridge had got past the foundation stage, and his assistant, George Daniel Stevenson, took the reins, making ornamental changes which included cladding the structure in its distinctive granite cloak instead of the red brick which had originally been planned. (Although not clearly visible, around 30 million bricks were used in building the bridge and its approach roads.)

FIG. 9.5: TOWER BRIDGE, LONDON: *The pivoted decks or bascules are still raised today despite the loss of shipping. A timetable of openings can be found on site or at www.towerbridge.org.uk.*

Building work had begun in the previous year with the establishment of two huge piers in the river which would take the load of up to 70,000 tonnes from the towers, bascules, suspended side spans, and walkways. The task was complicated by the insistence of the authorities that the building of the bridge should not interfere with traffic and a clearance width of 160 ft be kept open at all times. As a result, only one pier could be built at a time and not with the usual single caisson used in similar schemes. A unique arrangement of individual open wrought-iron caissons sunk in a ring to form the approximate outline of the pier was used, and the concrete and masonry were built up inside. The piers were made more complicated as a cavity had to be incorporated to house the counterbalance arms of the bascules as well as the machinery to move them.

Although it is assumed to be built of stone, the structure of the bridge is made of steel with the towers formed from a metal frame which, together with the suspension trusses and tie rods, takes all the load. The stone cladding is just that and serves no structural purpose. The towers consist of four octagonal columns set on large slabs of granite on top of the piers, through which they are bolted. They are held together by girders which form the internal landings and cross-bracing designed to resist a similar wind pressure to the Forth Rail Bridge (people were still concerned about the Tay Bridge disaster the decade before). As discovered on the Forth, steel is prone to corrosion, and so the sections of the columns which would be hidden by the outer cladding were coated with layers of oil canvas and cement.

The high-level walkways were cantilevered out from the completed towers a short way and then a suspended central section was fitted between (you can just see the white suspension members of this against the blue of the walkway). The walkway was not only designed to permit pedestrians to cross when the bascules

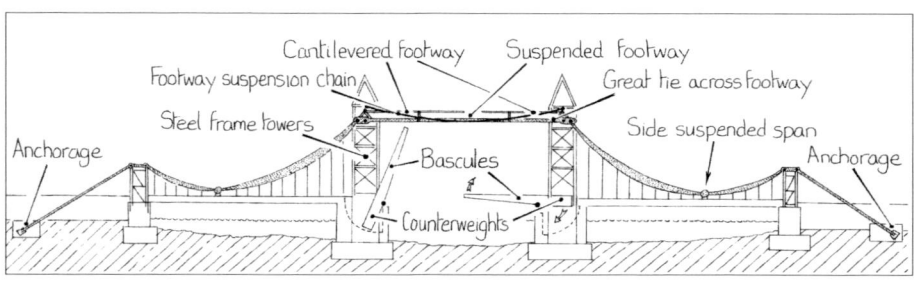

**FIG. 9.6:** *Despite being catalogued as a moving bridge, the main part of the structure is in fact a suspension bridge, as this diagram with all the cladding removed shows. The side spans are supported from catenary arched trusses anchored at each end and connected across the river by a great tie built into the walkway.*

# Moving Bridges

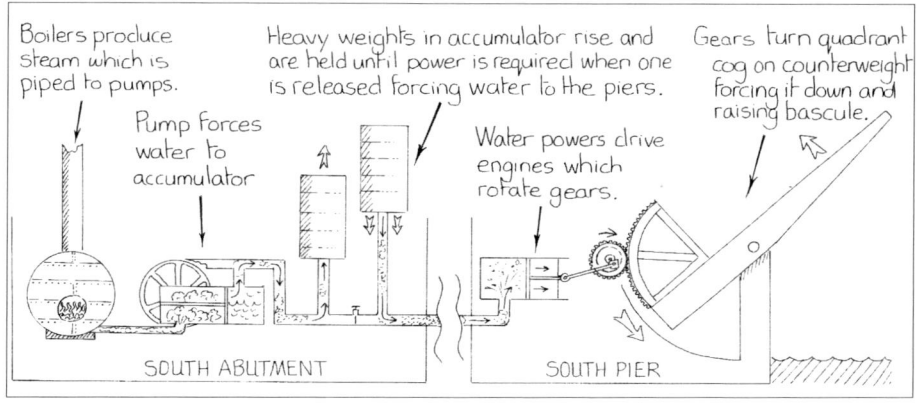

**FIG. 9.7:** *Diagram showing the hydraulic system which powered the lifting engines. The system was devised and supplied by Sir William Armstrong of Newcastle (see pages 121–122). Boilers provided steam to pumping engines, which forced up two huge cylinders, called accumulators, keeping a permanent pressurized water supply on tap. When the controls were thrown, the accumulator fell, pushing water down the pipes to power the engines which turned the cogs slotted in the rack on the back of the bascules.*

were raised – as it turned out, few used it, as they preferred to watch the spectacle below, and the walkway was closed in 1910 – but also included 300 ft long tie rods which linked the two ends of the side span suspension trusses together. If there was no connecting rod between them the side spans would simply pull the towers over. The walkway also carried the hydraulic pipes from the south to the north pier.

The most important part of the design was the lifting central section. When the bridge was down it was held in position by locks and rams, and when it had to be raised these were released, so that only a relatively small pressure needed to be applied to the large cogs which pushed the ends of the counterweights down. The power came from hydraulics, water stored under pressure in huge accumulators on the southern abutment, which, when the controls were turned, was released down the pipes to power the engines in each pier.

The finishing touch was to clad the the towers in their Gothic cloak blending French and Scottish baronial details which were popular at the time. The exposed metalwork, which is in a striking red, white and blue colour scheme at present, was originally brown, which can still be seen on the steel girders in the top of the towers today.

In some ways Tower Bridge marked the end of one era and the beginning of another. The cladding with which the Victorians had blanketed the frightful

# Bridges Explained: Viaducts & Aqueducts

FIG. 9.8: *This tall building which goes unnoticed at the southern end of the bridge houses the accumulators and the chimney for the boilers.*

new technology within came to be viewed with disdain in architectural circles. Increasingly a new breed of designers, especially those in the Arts and Crafts movement, preached honesty in building, although ironically the most honest of these new structures, the Forth Rail Bridge, was lambasted by the movement's leader, William Morris. It was Bazalgette's rejected steel arch suspended roadway design which would be used in similar locations in cities like Newcastle Upon Tyne, New York, and most famously at the harbour in Sydney. Nevertheless, Tower Bridge represented a new undercurrent of appreciation and support for the preservation of ancient buildings; this was the time when the National Trust and the Society for the Preservation of Ancient Buildings were formed. It was not just the most practical solution which was chosen but the one which best complemented the Tower of London, an aesthetic appreciation which would continue to affect designs in sensitive areas to the present day.

## SWING BRIDGES

Despite the success of Tower Bridge, large lift bridges were rare in this country. A far more common type of movable bridge across waterways was the swing bridge. This simple device is made by balancing a long deck on a central pivot so that it would swing horizontally out of the way of river or canal traffic.

The deck is supported on timber beams or metal girders in smaller examples and by more substantial trusses, above as well as below, on the largest types. The central pivot has a large ring around it with bearings or rollers for the structure to turn on, while the end of the deck has a shallow curved profile, matching a similar one on the landward side so that it does not clash when opening.

Although structures which swung or, in some cases, were pushed or pulled on sliders, were not new, it was as a result of the arrival of the canals and the demand for small, cheap movable bridges that they first became widespread. Canal swing bridges were built on an abutment of land narrowing the waterway so that the gap to cross gave just enough room for a boat to pass. The pivot could then be

# Moving Bridges

positioned on the edge of this abutment with half of the deck over the canal and the other half on land, where it could be pushed or pulled to open and close (see Fig 9.9). Many of these bridges have long since gone but the remains of the abutments and the sunken plane in which they swung can often be seen.

## Newcastle Swing Bridge

By the mid-19th century, larger structures were required to cross increasingly busy major rivers and larger canals. One such example was in Newcastle, where the 18th-century stone bridge blocked the passage of ships up the Tyne. This was a problem for William Armstrong, whose armaments factory was further upstream, beyond the reach of the larger craft he wanted to fit out. His solution was to replace the old structure with a new swing bridge which could allow them to pass when required.

However, a bridge of this size brought with it new problems. The river was too wide for a swing bridge of the kind used on canals to work, and so the central pivot would have to be on a large pier in the middle of the water, a much trickier construction than building on land. Secondly, because of its weight, the bridge would now have to be turned by an engine and this too would have to be mounted on the pier. The final and perhaps the greatest problem was the source of power. As the bridge would have to open and close at irregular times and at short notice, a steam engine would have to be kept in steam at all times, a very expensive option.

But Armstrong had already championed a new form of power at his country residence, Cragside, one which could move machinery at the turn of a tap: hydraulics. In this case the water from a reservoir high above the house kept the piped water under pressure, but at the bridge in the centre of the city this was not available and so Armstrong had to invent a way of artificially creating pressurized water on a limited site. His answer was the accumulator, a tall cylinder with a weighted ram on top, which, as a steam engine pumped water into it, kept it under a pressure of around 600 p.s.i. It was of such a size that it had sufficient water to allow the bridge to open numerous times and so the steam

FIG. 9.9: *A simple canal swing-bridge which is pushed on its landward end to open and close. This requires only a small amount of pressure, as the half of the deck over the land counterweights the other half hanging over the canal.*

# Bridges Explained: Viaducts & Aqueducts

**FIG. 9.10: NEWCASTLE SWING BRIDGE:** *A view along the deck, showing the control cockpit raised above it to give a clear view of both river and road.*

view, which added mystery to the mode of operation.

Construction began in 1868, the bridge finally opening to traffic eight years later. The 281 ft long deck is controlled from a cockpit, set above to give a clear view of road and river traffic. A bridge this size has to be secured when closed so as not to be moved by the wind or a live load. Rams are used to elevate it slightly before the surprisingly small hydraulic engine turns the cogs and gearing to open it. The original engine still opens the bridge today after some 300,000 operations.

## Barton Swing Aqueduct

Numerous swing bridges of this type were built to carry roads across some of our largest rivers, but, when Sir Edward Leader Williams designed such a structure above the Manchester Ship Canal which he was in the process of planning, he faced a new and far greater

engine pump didn't have to constantly run to refill it. As it was not practical to build the accumulator on land because of the problem of running pipes over the river, the 60 ft cylinder was built into the base of the pier, hidden from

**FIG. 9.11: NEWCASTLE SWING BRIDGE:** *The huge deck gains its strength from the iron trusses along each side. The accumulator is hidden in the bottom of the stone pier.*

# Moving Bridges

**FIG. 9.12: BARTON ROAD SWING BRIDGE:** *Built in the 1890s across the new Manchester Ship Canal, this large swing bridge was similar to that built 20 years earlier at Newcastle. Adjacent to it was a larger structure which posed a much greater engineering problem: a swing bridge which would have to carry a canal.*

problem. What was being carried across his new waterway was not a road but a canal.

The Bridgewater Canal was one of the first built in this country and is credited with bringing about a revolution in transport. It was engineered by James Brindley, and its most notable feature was a three-arched stone aqueduct across the River Irwell, the first major one built in this country, and a popular tourist attraction. When the Manchester Ship Canal was being built in the 1890s it followed the river for part of its course and the old aqueduct had to be replaced by a swing bridge in order to give clearance for tall shipping. (An arch from the original structure is set into the abutments of a neighbouring road bridge, see Fig 11.3.)

Leader Williams designed the bridge as a large trussed trough resting on a central pivot. Next, though, he had to solve a number of problems to make this structure work. As water was precious on the old canal, and the time needed to empty it and fill it again each time it had to be swung was too great, he was told that it had to work full of water. This made the trough nearly 1,600 tons, and so he fitted hydraulic rams in the pier to take some of the strain by raising the trough slightly before it was turned. This meant the hydraulic machinery had to turn only around half that weight. (The accumulators were located in the central tower to power both the road

**FIG. 9.13: BARTON SWING AQUEDUCT, MANCHESTER:** *This large trussed trough was pivoted on the central island on the left, which also supported the accumulator tower which powered both bridges.*

# Bridges Explained: Viaducts & Aqueducts

and canal swing bridges.) The bridge would have to be perfectly horizontal as it swung so that the water in the trough would not start to slosh around and create movement which could upturn the structure. A huge 25 ft metal ring with 64 rollers, each more than a foot in diameter, ensured a smooth passage. Before the engines were engaged, however, the ends of both the bridge and the canal would have to be sealed. This was done by means of gates, which were pivoted to one side and swung out, powered by hydraulic engines, the pressure of the water behind them helping to hold them tightly shut while the bridge moved.

What turned out to be the greatest challenge was sealing the gap between the canal and bridge when it was back in place. The bridge could not be an exact fit between the abutments, as a gap had to be left for expansion and for the trough to open without jamming. The gap would obviously leak water out as soon as the gates were swung open after the bridge was closed, and so Leader Williams designed a giant 12 ton 'U' shaped wedge lined with India rubber, to be lifted up when the gates were shut tight and then rammed back down into place before they were reopened. Most swing bridges have a shallow curved end but this would have made the wedge an awkward shape, and so the Barton Swing Aqueduct has a slanted end, so that the corners of the trough won't clash with the bank but the wedge can still be straight. Although modified with electrics in the late 1930s, the bridge still functions as when it opened in 1893.

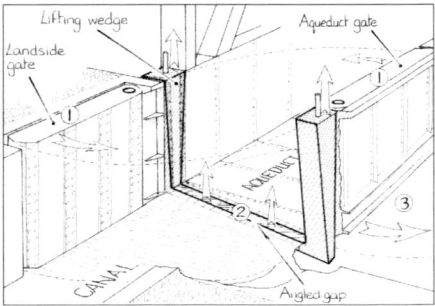

**FIG. 9.14:** *A view of the southern end of the bridge as it is today and a diagram below showing the same part (without the water). When opening the bridge, the gates were first swung shut (1); the watertight wedge was then raised up (2); and finally the trough was rotated (3). Note the angled gap between the bridge and the land, designed to avoid the bridge colliding with the land when it opened.*

## TRANSPORTER BRIDGES

For a brief period in the early 20th century, another solution to the problem of bridging a busy waterway became popular: the transporter bridge. It was less a bridge and more an aerial ferry, but, unlike a boat, was not hindered by tides, river flow, and weather; and it was quicker and less of a hindrance to river traffic than other bridges. The principle was that a cradle was carried over the river on cables hung from an overhead gantry and powered by motors which moved it back and forth. The massive trussed structure from which it hung could be cantilevered or suspended by cables.

Only four were ever built in this country. The largest, over the Mersey between Runcorn and Widnes, was demolished in 1961 (at greater cost than to originally build it). A privately owned example, at a chemical works in Warrington, ceased operations in 1967, although it still stands. The remaining two have stood the test of time and are operating today, creating distinctive landmarks at Newport, Gwent and Middlesbrough, Teeside.

## Newport Transporter Bridge

The treacherous River Usk formed a barrier between the two parts of Newport, especially to the south, where its swift flow and exceptional tides meant even a ferry crossing was too dangerous (an attempt was made but ended after fatalities). With plans in place to open a new steel works nearby, the need for a permanent crossing became imperative. After the rejection of other schemes on the grounds of cost, a transporter bridge, then a relatively new concept, championed on the continent by the French engineer Ferdinand Arnodin, was seen as ideal. He had pioneered new forms of suspension bridges using a combination of stays and vertical cables to support a stiffer truss, and this technique was used to support the gondola on the Newport Bridge.

Caissons were used to build the cylindrical piers, with the foundations reaching down some 80 ft, while timber piles were driven in to make a secure

**FIG. 9.15: NEWPORT TRANSPORTER BRIDGE, GWENT:** *The gondola is suspended from a moving platform which runs along the underside of the horizontal boom. This is supported by suspension cables which run over the top of the towers and are then anchored in the same way as a conventional suspension bridge.*

base for the two anchorages. The steel towers were then built up to 240 ft; saddles were fitted on top and the 16 main cables spun across (spirally wound rather than straight as on modern suspension bridges). The metal boom could then be built out, supported close to the towers by a fan arrangement of cable stays and then in the central portion by the main suspension cable. The gondola, which today can carry up to six vehicles and 120 passengers, was hung by cables from a long travelling platform on wheels, running over the lower girder of the boom, which spread the load over a longer section and lessened the pressure on any one part.

Opened in 1906, it never repaid the initial bill for construction and it cost more to operate than it brought in from tolls. This may explain why the maintenance stipulated by Arnodin was never carried out to the required standard. It closed in 1985 because of structural defects, but strong local support and government grants meant that it was repaired and put back into service in 1995, making this the longest surviving transporter bridge and a striking landmark and tourist attraction.

## Middlesbrough Transporter Bridge

The rapid growth of Middlesbrough led Charles Smith, the manager of an ironworks, to put forward a plan for a bridge in 1873. This would have been no ordinary structure: in fact it would have been the first transporter bridge in the world, except that the council decided to stick with a ferry and it was not built. It was more than 40 years before the idea was suggested again, and after delays for debate the contract was eventually signed in 1909.

Although it appears to have the same form as the Newport bridge, the

**FIG. 9.16:** *(Top) close-up of the suspension cables running into the anchorage; (centre) the gondola parked under the tower; (bottom) the stone cylindrical piers supporting the base of the tower.*

# Moving Bridges

**FIG. 9.17: MIDDLESBROUGH TRANSPORTER BRIDGE:** *Opened in 1911, it has remained an important link to this very day, and is the only example in continuous use in this country.*

## MODERN MOVING BRIDGES

Although commercial river traffic has virtually ceased inland, any new bridges built across large waterways still have to accommodate it. Redevelopment in Manchester has seen a number of inventive lift bridges built across the old ship canal. Ingenious designs for small footbridges include a rolling bridge in Paddington Basin, London, which curls up vertically to clear the canal.

The most famous by far is the Gateshead Millennium Bridge, or the 'Blinking Eye', as it is more

**FIG. 9.18: GATESHEAD MILLENNIUM BRIDGE:** *Close-up of the pivot and rams which move the counterbalanced structure to create the 'blinking' movement.*

method of supporting the main boom is completely different. Rather than a system of suspension cables, it used the cantilever principle, as successfully demonstrated by the Forth Rail Bridge a decade before. The 570 ft horizontal truss was built out from each side to meet in the middle. At the Forth the end piers of the approach viaducts were built as heavily weighted towers, in part to counter the weight of the live load; at Middlesbrough vertical cables running down from the overhanging ends to anchors in the ground carried out a similar task. The extra strength and stability of this type of bridge helped to enable the electrically propelled gondola to hold a greater weight – up to 12 vehicles and 600 passengers – which may have helped to make this still a viable operation in these days of vastly increased traffic.

affectionately known. It consists of two parabolic arches like open eyelids with cables between them. The upper one supports the lower footbridge and cycleway and counterbalances the bridge when it opens. This is done simply by pivoting the whole structure by hydraulic rams set in the base at each end, giving a clearance for boats of over 80 ft and creating the winking movement, which is proving a popular tourist attraction (the opening times are published in the glass buildings at each end of the bridge).

FIG. 9.19: GATESHEAD MILLENNIUM BRIDGE: *The bridge in its open position (left) and closed position (bottom).*

# Chapter 10

# *Viaducts*

**FIG. 10.1: RIBBLEHEAD VIADUCT, NORTH YORKS:** *Its daunting size and spectacular location make this one of the best-known viaducts in the country. It was designed by John Sydney Crossley and took nearly four years to build. Because of the boggy ground (after which it was originally named) some of the piers had to be sunk down 25 ft before a good footing could be established. Nearly 1,000 navvies were working on this part of the line and had to be accommodated in shanty towns, some with pubs, shops, a library, and a hospital; the grassed-over footings of these prefabricated buildings can still be seen nearby.*

## The Development of Viaducts

A viaduct is a series of roughly equal spans across a valley. The examples in this book are used to carry railways and roads, and comprise tall piers with the gaps spanned by trusses, girders, or, most commonly, arches, all being of roughly the same form and size. Large multi-span structures which may look the same as a viaduct are usually referred to as bridges if they cross just a river, and they may

# Bridges Explained: Viaducts & Aqueducts

have a wider span in the middle for navigation.

It may be expected that the most magnificent viaducts were built later in the railway age as engineers grew in confidence but the reverse was often the case. Early steam locomotives had poor traction compared with later types, and so steep gradients were avoided on the first railway lines. This means that more tunnels and viaducts were required and some of the largest date from the earlier phase of railway construction, from the 1830s to the 1850s. Most were masonry or brick arched structures, typically with a semi-circular form so as to convey most of the load downwards rather than sideways onto the tall piers, although some were built with segmental or even elliptical arches.

As with all bridges the foundations were very important and, although viaducts mainly crossed dry land, many had to have extra piling or even filled sheepskins underneath to give a sound footing over boggy land. The piers were built up in scaffolding, with either sockets or corbels (small brackets) fitted at the top to support the timber centring on which the arches were built. If the construction was not revolutionary in the methods used, then the scale of the operation certainly was challenging. Stockport Viaduct was typical of many, using an incredible 31 million bricks in its construction.

Brunel was notable among engineers for building a large number of inventive timber truss viaducts on his railways across the south-west. His networks of beams resting upon short stone or brick

**FIG. 10.2: NORTH UNION RAILWAY BRIDGE, PRESTON:** *Although it appears to be a viaduct, this multi-arched structure is called a bridge as it crosses a river only.*

piers were considerably cheaper than the alternatives until the price of the Baltic timber he used shot up. It will never be known how long they would have lasted, as all of them have been demolished, replaced, or buried in embankments (see Fig 1.9). Later some viaducts were built using iron piers and trusses to keep the cost down, most impressively at Crumlin in South Wales, although, as is typical with much of the finest Victorian engineering, it has since been demolished. Meldon Viaduct is the only notable example left of this type (Fig 10.7).

In the second half of the 19th century, just as in bridge construction, iron lattice trusses were being used for some viaducts. Many were still built with masonry and brick arches, including the longest in the country

# Viaducts

**FIG. 10.3: COALBROOKDALE, SALOP:** *Although most viaducts had semi-circular arches, there were some built with segmental arches, as in this brick example.*

London North Eastern's lines into Scotland, but a dispute in the 1860s forced them to plan a railway of their own. The route they chose from Settle to Carlisle crossed some of the most remote and hilly countryside in England; yet, as it was going to be a main line, the track had to avoid tight turns and have no gradients steeper than 1 in 100. To achieve this numerous tunnels and viaducts would have to be built, a very expensive option and so, when, after the passing of the government Act in 1867, booming interest rates created a volatile market, the Midland decided to abandon their plans. Unfortunately for them there were a number of other interested parties, especially smaller railway companies, who would benefit from this new main line and they petitioned the government to enforce the

(outside of London) across the Welland Valley, with an amazing 82 arches (Fig 10.6 and 10.10). Increasingly towards the end of the century, new, very strong engineering bricks were used, typically blue or grey in colour, which help to date these later structures, although they were sometimes used for repairs or extensions to earlier examples (see Fig 10.11). A number of viaducts were built of concrete, not reinforced in beams or arches, but as solid blocks simply replacing stone in conventional arched structures (see Fig 10.12).

## Ribblehead Viaduct

Arguably our most famous railway line was very nearly never built. The Midland Railway had been sharing the

**FIG. 10.4: CHAPEL-EN-LE-FRITH, DERBYS:** *Many viaducts have retained the corbels at the springing of the arch which supported the centring while they were being constructed.*

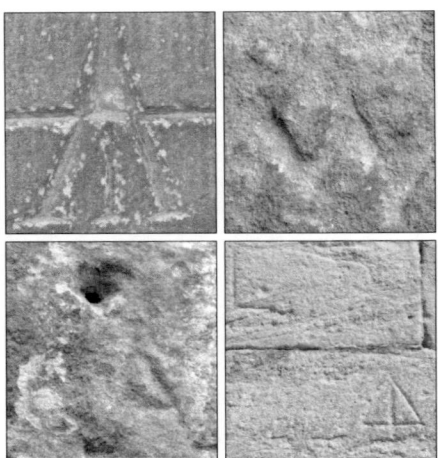

FIG. 10.5: *Many stone bridges and viaducts have strange marks in the masonry. These are usually the signature left by the masons working on the project. The hole in the bottom left example was used to lift the block into position.*

FIG. 10.6: WELLAND VALLEY VIADUCT: NORTHANTS: *With 82 arches, the longest viaduct outside of London, it was built in 1876 in engineering bricks, which was the popular material of the day.*

Act and in November 1869 the Midland had to reluctantly concede and commence construction.

The most impressive feature of the line, and one of the trickiest to build, was the Ribblehead Viaduct, a huge masonry structure over 100 ft tall, comprising 24 arches built over boggy ground on high moorland. So remote was the site that a shanty town was built nearby for the navvies, but the isolated life coupled with disease and numerous accidents caused such a loss of life that the railway company had to pay for an extension to the local graveyard.

FIG. 10.7: MELDON VIADUCT, DEVON: *A unique wrought-iron trussed viaduct, built in 1871.*

# Viaducts

**FIG. 10.8: WHARNCLIFFE VIADUCT, HANWELL, LONDON:** *Brunel's first major structure along the Great Western Railway, built in 1837. It has large elliptical arches similar to Maidenhead Bridge (Fig. 4.28). Named after Lord Wharncliffe, who helped the railway's bill pass through Parliament.*

**FIG. 10.9: BALCOMBE VIADUCT, WEST SUSSEX:** *1,450 ft long brick viaduct with decorative stone parapets and, at each end, an Italianate pavilion. This imposing structure was typical of many early viaducts and was completed in 1841. The lozenge-shaped holes in the piers were to lighten the structure, which used 11 million bricks, shipped over from Holland.*

**FIG. 10.10: WELLAND VALLEY VIADUCT, NORTHANTS:** *This colossal viaduct was designed by John Underwood, who, as was often the case, was the railway company's own engineer.*

# Bridges Explained: Viaducts & Aqueducts

FIG. 10.11: MARPLE, CHESHIRE: *Many railways followed the course of canals and it is not unusual to find a viaduct and aqueduct side by side across the same river, as here at Marple. The engineering bricks on the right arch are from a later repair.*

FIG. 10.12: CANNINGTON VIADUCT, DORSET: *A rare concrete viaduct, completed in 1904 using blocks cast on site and mass concrete infill. The corbels which supported the centring can still be seen under the arches. Settlement in the foundations at one end meant a jack arch had to be inserted during building (the double arch at the far end).*

FIG. 10.13: BICKLEIGH, DEVON: *Brunel built a series of timber tressle or truss viaducts resting on stone piers along the Great Western line from Tavistock to Plymouth. These were replaced by stone and brick structures in the 1890s, but the original piers can still be seen behind the later viaduct in this example.*

# Viaducts

**FIG. 10.15: DIGSWELL, HERTS:** *This huge brick viaduct of 40 spans was built by Lewis Cubitt for the Great Northern Railway and completed in 1850. At its highest point it is over 100 ft above the valley below.*

**FIG. 10.14: CHIRK VIADUCT, CLWYD:** *A 100 ft high stone viaduct completed in 1848 by Henry Robertson. Note the decorative niche and dentil feature along the top.*

**FIG. 10.16 (below): CHAPEL-EN-LE-FRITH, DERBYS:** *Two early stone viaducts.*

# BRIDGES EXPLAINED: VIADUCTS & AQUEDUCTS

**FIG. 10.17: CONGLETON VIADUCT, CHESHIRE:** *Although built over 150 years ago for the North Staffs Railway, this large brick structure over the River Dane valley still carries high speed trains.*

**FIG. 10.18: LOCKWOOD VIADUCT, HUDDERSFIELD:** *Completed in 1849, it carries the railway 120 ft above the valley on 32 arches. The stone used was dug out of the cuttings on each approach to the viaduct, a total of nearly 1 million cubic feet of masonry.*

## Motorway Viaducts

Viaducts are still built to carry motorways across large valleys or complicated intersections. They are generally reinforced concrete columns or piers with box girders or steel plate girders carrying the concrete road deck.

**FIG. 10.19** *(left)*: **THELWALL VIADUCT, WARRINGTON:** *This famous traffic bulletin location carries the M6 over the Mersey valley between junctions 20 and 21. It is made from huge steel-plate girders resting on transverse concrete beams on top of columns. Although its double structure is referred to as a viaduct, the earlier western section (built in 1963) has a wider, cantilevered span across the main river.*

## Chapter 11

# *Aqueducts*

**FIG. 11.1: PONTSYCYLLTE AQUEDUCT, CLWYD:** *This imposing aqueduct across the River Dee near Wrexham was built by Thomas Telford and completed in 1805. Its breathtaking elegant form is heightened by its stunning location across a deep wooded valley.*

## The Development of Aqueducts

An aqueduct is a channel supported on piers or abutments used for carrying water. Virtually any size of structure carrying water, from a single span bridge to one standing upon numerous piers, is referred to as an aqueduct. They have to carry the permanent weight of the water held within, which exerts a downward and sideways pressure. This tends to be the major factor in their design. The Romans were the first to build aqueducts in this country, although never on the scale of the spectacular multi-arched structures

# Bridges Explained: Viaducts & Aqueducts

on the continent. They were used only for water supply, as were numerous small scale structures built since.

It was in the mid 1750s when aqueducts to carry boats began to be built in this country. The first to catch the public's imagination was designed by James Brindley to cross the River Irwell at Barton, Manchester, on the Bridgewater Canal. There were howls of derision at the time, as some thought he was crazy to carry a waterway over a river on an arched structure. It was described by one observer as the castle in the air. To Brindley it was not so much the arches which caused a problem as the means to contain the canal within. It was an unknown science at the time, and so Brindley designed the Barton Aqueduct with thick sides to the channel, a batter to the outer wall, sloping outwards at the bottom for stability, and buttressed piers to resist the outward force of the water. The narrow navigation channel was lined with puddle, a watertight clay mixture.

Aqueduct design made dramatic improvements in a very short time, and the next generation of masonry

**FIG. 11.2: LONGDON ON TERN AQUEDUCT, SALOP:** *This humble 62 ft span supported upon cast-iron piers was a revolution when built and still stands. Although a similar iron structure was opened a few months earlier, this one, built completely in iron, is generally regarded as the first of its type in the world. Telford had to overcome the problem of how to waterproof the joints between the plates, the solution for which he found to be a mixture of flannel soaked in boiling sugar and sealed in with lead. He also found that the narrow channel, which was barely wide enough for a boat, caused problems, as the water needing to pass behind it as it moved forward could not get past the hull quickly enough, meaning that towing the boat was hard and slow work. This was something he would put right on the Pontcysyllte Aqueduct. The right-hand photo shows inside the drained aqueduct, with the towpath on the left and the channel on the right.*

# Aqueducts

aeducts were often more ambitious and elegant in form. To reduce the overall bulk of the structures, iron tie-rods were run through to help resist the outward force so that the surrounding infill could be reduced in size, making savings on the materials and weight of the finished aqueduct.

## Lune Aqueduct

John Rennie, a contemporary of Telford, was one of the leading canal engineers and bridge designers of the late 18th and early 19th centuries, who built two of the most important bridges over the Thames: Waterloo and London, both now replaced. Some of his early work was on the Kennet and Avon Canal, where he built two aqueducts of note over the River Avon. At the same time he was working on the Lancaster Canal, an ambitious scheme to link the northern towns of Preston, Lancaster, and Kendal with the main waterways network to the south. This would require two large aqueducts crossing the rivers Ribble and Lune.

One of the first parts to be built was the Lune Aqueduct, a massive five-arched structure, over 600 ft long, carrying the new canal over the river at a height of around 60 ft. Its construction followed that of traditional arched bridges: cofferdams were formed and timber piles driven into the river bed as a foundation. Hollow piers were built up and then filled with rubble, and centring for the semi-circular arches was erected. The difference was the channel built inside to carry the waterway. Here, the internal stone sides and base were

FIG. 11.3: BRINDLEY'S RIVER IRWELL AQUEDUCT ARCH: *The Barton Aqueduct over the River Irwell, completed in 1761, was the first major structure of the canal age, totalling 600 ft in length with its approaches. The duke whose idea it was to build the waterway had planned to use locks down the river, but he adopted the more ambitious plan of an aqueduct proposed by James Brindley, despite his having no experience of building canals (he had been a wheelwright who had his own mill at Leek in Staffordshire). It carried the Bridgewater Canal on three arches, 38 ft above the river, with an overall width of 36 ft and an 18 ft wide channel. The large, solid sides were battered (leaning out slightly) and, together with buttresses, were used to hold the water pressure in check. It was replaced by the Barton Swing Aqueduct in the 1890s when the Manchester Ship Canal was built (see Fig. 9.13). An arch which was part of the original structure crossing the lane which ran adjacent to the river was saved and built into a road abutment a few hundred yards to the north.*

**FIG. 11.4: RIVER DOVE AQUEDUCT, BURTON UPON TRENT, STAFFS:** *Another early aqueduct designed by James Brindley on the Trent and Mersey Canal, near Burton upon Trent. It has a massive body to hold the channel, with clay lining between the inner and outer walls. The channel width of 28 ft (greater than at Barton) permits two boats to pass. The 12-arch crossing of the River Dove is typical of early aqueducts: low segmental arches 15 ft wide but only 4 ft high.*

**FIG. 11.6: DUNDAS AQUEDUCT, BATH:** *This stone aqueduct by John Rennie had a 64 ft segmental arch with two narrower flanking arches to help relieve the pressure on the structure at times of flood. However, it is the restrained classical ornamentation which captures the imagination.*

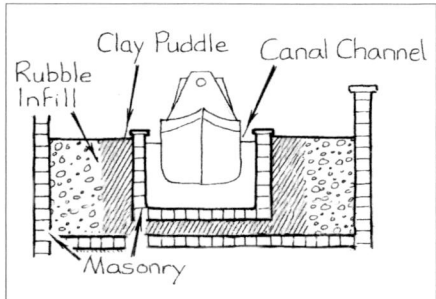

**FIG. 11.5:** *A cross-sectional diagram of an aqueduct showing how the interior may have looked. The infill to the sides could be entirely of clay puddle or a mix of materials as shown here.*

curved, supported by lateral inverted arches, with a thick puddle lining, and iron tie-rods running from one side to the other to help resist the outward pressure (the ends hidden behind the exterior stonework).

The form of this massive aqueduct is worthy of note, but it is the classical cornice (a horizontal decorative band near the top) and the balustraded parapet which make it such an elegant and outstanding structure. However, all this masonry came at a price – in fact one which was three times over budget – so that there was not enough money

# Aqueducts

to built the aqueduct over the Ribble. The northern and southern sections of the canal had to be connected by a cheaper tramway and were never linked as originally planned, although a connection was finally built further downstream, for the Millennium.

At the same time as these huge masonry structures were being built, Thomas Telford and William Jessop were taking a dramatically different direction. The iron trough designed by Telford at Longdon on Tern demonstrated how it could contain the water without any surrounding infill and masonry, making the load significantly lighter. This construction would permit a much taller structure, since it would not exert the weight and outward pressure on the piers as when masonry arches were used.

## Pontcysyllte and Chirk Aqueducts

Legend has it that Thomas Telford made his first noteworthy contribution in Shropshire when he was asked for his opinion on the state of St Chad's church roof in Shrewsbury. He surprised the authorities by telling them that it wasn't the roof they should be concerned about but the tower, which was in his opinion on the verge of collapse. His advice was ignored but only a few days later the tower promptly collapsed, destroying so much of the structure that it had to be

**FIG. 11.7: LUNE AQUEDUCT, LANCASTER:** *This graceful structure carrying the Lancaster Canal across the River Lune was designed by John Rennie and opened in 1797.*

# Bridges Explained: Viaducts & Aqueducts

FIG. 11.9: MARPLE AQUEDUCT, DERBYS: *Designed by James Outram and completed in 1800, this graceful masonry aqueduct carries the Peak Forest Canal over the River Goyt at over 100 ft. Its attractive form is in part due to the thin body of the crossing, as the channel is very narrow, brick-lined, and backed with clay puddle, and has thinner retaining sides than earlier aqueducts. Although this helps to keep the weight down, Outram also used the method adopted by William Edwards at Pontypridd (see Fig. 4.1) and built 12 ft diameter holes through the spandrels.*

FIG. 11.8: *Classical details from Lune Aqueduct. The inscription (top) on the upstream side is 'To Public Prosperity'. On the downstream face is a Latin phrase which roughly translates as 'Old needs are served, far distant sites combined. Rivers by art to bring new wealth are joined'.*

completely rebuilt on a new site. With his reputation established in the county, Telford proceeded with various tasks in his role as Surveyor of Public Works; he was also appointed assistant to Josiah Clowes, the chief engineer of the Shrewsbury Canal. Telford's career was to take another turn when Clowes died with the project only half complete and the brick and masonry aqueduct crossing the River Tern at Longdon washed away. Telford took over and despite wanting to rebuild this awkward structure in stone was forced to use iron by one of the canal backers, a local

# Aqueducts

iron-master called William Reynolds. Telford's inventive solution was to build a metal trough formed from cast-iron plates shaped like the voussoirs of an arch and bolted together. Although this was one of the first of its kind in the world, it was to be more important to Telford in the long run as a prototype for something he was soon planning which would be much bigger, fifteen times longer, in fact: the Pontcysyllte Aqueduct.

Before he started his work on the Longdon on Tern Aqueduct, Telford had been appointed to work under the great canal engineer William Jessop on the design and construction of the Ellesmere Canal, to link the Mersey to the Severn. It was to run through Chester, up to the iron and coal fields to the west of Wrexham and then down to Shrewsbury, with various branches connecting it to valuable markets. This huge undertaking received its Act in 1793 and various parts of the network were completed in a few years before attention was directed to the major crossings of the River Ceiriog at Chirk and the River Dee near Froncysyllte (the aqueduct was named after the earlier three-arched bridge in the valley below, hence the prefix Pont).

Originally it had been planned that the canal would go down into the valley and up again the other side by means of locks, the easier solution but one which creates problems with water supply. The major headache for all artificial waterways is keeping water in them, and locks transfer huge quantities downstream each time a boat goes down. If every river is crossed by

**FIG. 11.10: CHIRK AQUEDUCT, CLWYD:** *The aqueduct to the left appears to be built completely of stone, but the trough which holds the water is cast iron, a cautious step towards Telford's next project: the Pontcysyllte Aqueduct (the viaduct on the right is in Fig. 10.14).*

this method, then the water in the upper section will quickly be drained away and huge pumping engines are required to return the lost water back to the top. If an aqueduct is built then the summit section is greatly increased in length and will hold a larger quantity of water. This means a lockful drained off at the end has less effect on the overall levels. So it was soon realized here that two aqueducts would greatly save operational and maintenance costs and the idea of using locks was dropped at an early stage.

The first of the aqueducts constructed was at Chirk and, perhaps reflecting the engineers' caution, it was built in masonry but with an iron base to the trough – metal sides or a metal

# Bridges Explained: Viaducts & Aqueducts

FIG. 11.11: *The soffit of one of the arches showing the square open access hole into the hollow interior and a line of sockets below, some with the timber ends of the centring still in place.*

lining appear to have been added at a later date – which not only helped to lighten the structure but also helped to tie in the two masonry sides. The piers were built with a solid interior up to the springing of the arches; above that point they were hollow, including the spandrels, with cross-walls built inside for strength. This not only lightened the structure but also permitted inspection of the iron base plates; the access holes on the underside are still open today.

When Chirk was opened in 1801, the Pontcysyllte Aqueduct was only beginning to take shape. Despite the foundation stone having been placed six years earlier, there had been many delays as the engineers debated the best solution, and funds were needed to complete other parts of the canal. Now, though, Telford and Jessop went ahead with their final plan: 19 iron troughs resting on the top of piers, each supported by four cast-iron segmental arches bracketed off the masonry. The huge piers were tapered, with a founding on rock and a solid infill in the first 70 ft and hollow above to reduce the weight (something which Telford tended to do with all his bridges). As at Longdon the troughs were cast like voussoirs, with external flanges bolted together. The same mix of flannel, boiling sugar, and lead was used to make the joints watertight. However, there was one significant change: rather than make a narrow channel with the towpath fixed on the side, he made the trough wider and cantilevered the towpath over the top of the water so that there was no repeat of the problem with water passing the boat hulls as at Longdon on Tern (see Fig 11.2).

FIG. 11.12: PONTCYSYLLTE AQUEDUCT, CLYWD: *A close-up of one of the troughs showing the bolted flanges which held each voussoir-shaped casting together and the four iron ribs which supported them.*

# Aqueducts

FIG. 11.13: PONTCYSYLLTE AQUEDUCT, CLYWD: *For all of Telford's genius in designing the aqueduct, the canal which it was built for was never completed and ended just a few hundred yards to the north.*

the towpath handrail. But this is nothing compared with crewing a narrowboat across, as, due to the low level of the cast-iron trough, the side of the boat is higher and, despite sockets having been cast into the trough for a railing on the canal side, one was never fitted! There is literally nothing to stop you simply stepping off the boat and plunging 120 ft to the valley bottom.

FIG. 11.14: *Cast-iron aqueducts continued to be used in the first half of the 19th century before canal building declined in the face of railway mania in the 1840s. The top example is at Stretton, Staffs; it was designed by Telford and built in 1832. The lower example is one of many built across the new Birmingham Canal main line in the 1820s and 30s.*

The completed structure was the longest and highest aqueduct built in Britain: over a 1,000 ft long and nearly 120 ft high. The embankment which leads up to it is also colossal but is lost in trees today. Its popularity is in part due to its elegant form, the dramatic setting high above the wooded Dee valley, and the connection with Telford. Yet to many of those who have crossed it, the lasting memory is of terror. For those of us who feel the knees wobble on the first rung of a ladder, walking across the aqueduct is performed with a sweaty palm firmly clenched around

# BRIDGES TO VISIT

There are literally thousands and thousands of bridges to discover all over the country, including many spectacular examples. Although the majority relate to canals, railways, and modern roads, there are still numerous bridges from previous centuries, some possibly 800 or 900 years old. The following list includes those mentioned in the book and a few others of personal interest. The grid reference for each bridge is listed and if you do not have a map of the area then try www.ordnancesurvey.co.uk/oswebsite/getamap. Enter the grid reference and the bridge should appear in the centre.

## BATH
**Pulteney Bridge, Bath** (ST 752/649) see Fig 4.14. Famous bridge with shops designed by Robert Adam.
**Dundas Aqueduct, Bath** (ST 785/625) see Fig 11.6. Ornate stone aqueduct over the Avon.
**Clifton Suspension Bridge** (ST 564/731) see Fig 8.12- 8.14. Brunel's iconic chain suspension bridge across the Avon gorge.

## BEDFORDSHIRE
**Great Barford Bridge** (TL 134/516) see Fig 4.2. Medieval bridge with a 19th-century Gothic brick façade.
**Sutton Packhorse Bridge** (TL 220/474): see Fig 2.41. Impressive two-arch packhorse bridge next to a ford.
**Bromham Bridge** (TL 011/506) see Fig 2.36. Medieval bridge with a large causeway.
**Harrold Bridge** (SP 955/565) see Fig 2.29. Rustic bridge with arches of differing size and form as each one had a different person responsible for its building and repair.
**Oakley Bridge** (TL 008/529) see Fig 1.6.

## BERKSHIRE
**Maidenhead Bridge:** (SU 901/814): see Fig 4.19. Attractive classical stone multi arch bridge across the Thames.
**Railway Bridge, Maidenhead** (SU 901/810): see Fig 4.28. Brunel's famous brick segmental arch bridge over the Thames.
**Sonning Bridge** (SU 755/757): see Fig 3.1. Red-brick bridge dating from the late 18th century, despite looking earlier.

## BUCKINGHAMSHIRE
**Marlow Suspension Bridge** (SU 851/861): see Fig 8.11. Picturesque early chain suspension bridge over the Thames.
**Thornborough Bridge:** (SP 729/332): see Figs 2.17, 2.30, 3.10. 14th-century pointed arches and upstream cutwaters.

FIG. 12.1: EAMONT BRIDGE, PENRITH, CUMBRIA

# BRIDGES TO VISIT

**Ouse Bridge, Stony Stratford** (SP 781/409) see Fig 4.17. Low segmental arched structure from 1835.
**Tyringham Hall Bridge** (SP 858/465) see Fig. 4.3. Classical styled single arch bridge designed by Sir John Soane.
**Palladian Bridge, Stowe** (SP 680/372) see Fig 4.12. Classical covered bridge from c. 1740.
**Tickford Bridge, Newport Pagnell** (SP 878/438) see Fig 5.12. Early cast-iron bridge.

## CAMBRIDGESHIRE
**Clare Bridge, The Backs, Cambridge** (TL 445/585): see Fig 4.8. Very early classical bridge.
**Huntingdon Bridge** (TL 243/714): see Figs 2.13, 2.17. Late 14th-century, with pointed arches, cutwaters, and refuges.
**St Ives Bridge** (TL 312/711): see Figs 2.2, 2.21.
**Nun's Bridge, Hinchingbrooke** (TL 226/712): see Fig 2.34. Medieval bridge with mix of round and pointed arches.

## CHESHIRE
**Old Dee Bridge, Chester** (SJ 407/657): See Fig 2.35. Late 14th-century sandstone multi-arch bridge.
**Grosvenor Bridge, Chester** (SJ 402/655): see Fig 4.20. Large 200 ft wide segmental arch bridge.
**Queen's Park Suspension Bridge, Chester** (SJ 896/657): see Fig 8.15.
**Congleton Viaduct, North Rode** (SJ 896/657): see Fig 10.17.
**Congleton Turnover Bridge** (SJ 869/622) see Fig 4.24. One of a number of turnover canal bridges along the Macclesfield Canal.

**Holt Bridge** (SJ 412/544) see Fig 2.23, 2.31. 14th-century sandstone bridge.
**Iron Bridge, Eaton Hall** (SJ 418/601) see Fig 5.14. Ornate iron bridge by Telford over the River Dee.
**Wrenbury Lift Bridge** (SJ 590/480) see Fig 9.3. One of a number of working lift bridges along the Llangollen Canal.
**Thelwall Viaduct** M6 Junc 20-21 (SJ 664/882) see Fig 10.19.
**Runcorn Bridge, Runcorn/Widnes** (SJ 509/834) see Fig 5.20. Huge steel arch with suspended roadway.
**Runcorn Railway Bridge, Runcorn/Widnes** (SJ 509/835) see Fig 6.14. Lattice girder bridge with decorative piers, running parallel to road bridge.

FIG. 12.2: ASHNESS BRIDGE, DERWENT WATER

# BRIDGES EXPLAINED: VIADUCTS & AQUEDUCTS

FIG. 12.3: STOCKLEY BRIDGE, SEATHWAITE, LAKE DISTRICT

## CORNWALL
**Wadebridge** (SW 991/724): 15th-century stone bridge, 400 ft long with 15 pointed arches.
**New Bridge, Gunnislake** (SX 433/722): 16th-century granite bridge along A390, a mile north of village.
**Royal Albert Bridge, Saltash** (SX 435/587): see Figs 6.1, 6.18, 6.19. Brunel's famous tied arch bridge over the Tamar.
**Tamar Bridge, Saltash** (SX 435/588): see Fig 8.17. Locally funded large suspension bridge.

## CUMBRIA
**Lanercost Bridge, Lanercost** (NY 553/633) see Fig 3.11. Large segmental arched bridge from 1724.
**Eamont Bridge, Penrith** (NY 522/288) See Fig 12.1. 15th-century stone arch bridge.
**Grange, Derwent Water** (NY 254/174)
**Ashness Bridge, Derwent Water** (NY 270/197) See Fig 12. 2.
**Devil's Bridge, Kirkby Lonsdale** (SD 616/782) see Fig 3.2. Stunning high-arched structure from the late 15th century.
**Stockley Bridge, Seathwaite** (NY 235/109) see Fig 12.3. Early Lakeland packhorse bridge, probably late 17th century.

## DERBYSHIRE
**Bakewell Bridge, Bakewell** (SK 219/686) see Fig 2.32. 15th-century pointed arch bridge.
**Chatsworth Park Bridge** (SK 257/702) see Fig 4.13. Classical style bridge with statue, dating from 1759.
**Sheepwash Bridge, Ashford in the Water** (SK 194/696) see Fig 3.6. Beautifully positioned medieval packhorse bridge.
**Kedleston Hall Bridge** (SK 313/407): Classical styled ornamental bridge in the grounds of Kedelston Hall designed by Robert Adam.
**Chapel Milton Viaducts, Chapel-en-le-Frith** (SK 055/818): see Figs 10.4, 10.16. Pair of converging stone viaducts.
**Railway Bridge, Friars Gate, Derby** (SK 347/364): see Fig 5.9. Stunning cast-ron decorative railway bridge.
**Litton Mill, Manifold Way** (SK 158/730) see Fig 4.27. Railway bridge along the Manifold Way. Close by you

# BRIDGES TO VISIT

can walk across the viaduct over the River Wye.
**Youlgreave Clapper Bridge** (SK 209/640) see Fig 6.3. Smaller clapper bridge over the River Bradford.
**Dove Aqueduct** (SK 268/269) see Fig 11.4. Early Brindley aqueduct.

## DEVON
**Postbridge Clapper Bridge** (SX 648/788) see Fig 6.4. Clapper bridge of unknown date but could be a couple of thousand years old.
**Bideford Long Bridge** (SS 455/264): see Fig 2.38. Over 650 ft long, 24 arches; dates from around 1315.
**Meldon Viaduct** (SX 564/923) see Fig 10.7. Unique trussed viaduct from 1871.
**Bickleigh Viaduct** (SX 525/627) see Fig 10.13. Remains of the piers from one of Brunel's timber trussed viaducts alongside replacement from 1890s.
**Rothern Bridge, Great Torrington** (SS 479/198) see Fig 12.4. Medieval stone arch bridge.

## DORSET
**Town Bridge, Sturminster Newton** (ST 785/135): see Fig 2.37. Attractive late medieval bridge.
**Cannington Viaduct, Combpyne** (SY 316/924): see Fig 10.12. Rare concrete block railway viaduct.

## COUNTY DURHAM
**Elvet Bridge, Durham** (NZ 276/421) see Fig 2.33. Very early bridge, possibly from the 12th century.
**Barnard Castle Bridge** (NZ 048/164) Impressive medieval bridge beneath the castle.

FIG. 12.4: ROTHERN BRIDGE, GREAT TORRINGTON

## GLOUCESTERSHIRE
**Mythe Bridge, Tewkesbury** (SO 889/337): Cast-iron bridge by Telford across the Severn.
**Over Bridge, Gloucester** (SO 817/196) see Fig 4.22. Notable single arch bridge by Telford.
**The First Severn Suspension Bridge, Aust** (ST 559/902) see Figs 8.18–8.20. Groundbreaking aerofoil deck suspension bridge; best viewed from Beachley, nr Chepstow.
**Second Severn Crossing, Severn Beach** (ST 512/865) see Figs 8.30, 8.31. Huge concrete girder and cable stayed bridge. Excellent visitor centre covering both bridges.

# Bridges Explained: Viaducts & Aqueducts

FIG. 12.5: ITCHEN BRIDGE, SOUTHAMPTON

## HAMPSHIRE
**Itchen Bridge, Southampton** (SU 433/112): see Fig 12.5. Concrete box girder bridge.

## HEREFORDSHIRE
**Wye Bridge, Hereford** (SO 508/396): see Fig 1.14.
**Wilton Bridge, Ross-on-Wye** (SO 589/242) see Fig 3.14. Fine sandstone bridge from 1597, with unusual 18th-century sundial in centre.

## HERTFORDSHIRE
**Digswell Viaduct** (TL 245/149) see Fig 10.15. Large early railway viaduct.

## KENT
**East Farleigh Bridge** (TQ 734/535): 15th-century stone pointed arch bridge across the Medway.
**Kingsferry Lift Bridge** (TQ 914/693): Horizontally lifting bridge from 1960 linking the Isle of Sheppey with the mainland.

**Queen Elizabeth II Bridge, Dartford, M25** (TQ 570/764): see Fig 8.29. Cable stayed bridge carrying the M25 over the Thames. No easy access; best approached along Thames Path.

## LANCASHIRE
**Packhorse Bridge, Wycoller** (SD 932/392): see Fig 2.40. Wonderfully picturesque wonky packhorse bridge.
**Lune Aqueduct, Lancaster** (SD 484/639): see Figs 11.7, 11.8. Notable classically styled stone aqueduct.
**Skerton Bridge, Lancaster** (SD 479/623): see Fig 4.16. Large classical stone bridge with level roadway.
**Millennium Footbridge, Lancaster** (SD 476/621): see Fig 8.28.
**North Union Railway Bridge, Preston** (SD 535/283): see Fig 10.2.
**Tramway Bridge, Preston** (SD 542/287): see Fig 12.6. Unique timber truss bridge rebuilt in concrete.

## LEICESTERSHIRE
**Medbourne Bridge** (SP 799/931) see Fig 2.15. 13th-century packhorse bridge with wooden handrail.

FIG. 12.6: TRAMWAY BRIDGE, PRESTON

# BRIDGES TO VISIT

## LINCOLNSHIRE
**High Bridge, Lincoln** (SK 975/712) see Fig 2.24. Medieval bridge with 16th-century buildings.
**West Rasen Packhorse Bridge** (TF 062/893): see Fig 2.42. Tiny but original packhorse bridge.
**Humber Bridge, Barton upon Humber** (TA 024/240): see Figs 8.21–8.23. Largest single span bridge in Britain.
**Trinity Bridge, Crowland** (TF 239/102) see Fig 2.18. Unusual three-legged medieval bridge in the middle of the village.

## LONDON
**Tower Bridge** (TQ 336/802): see Figs 9.1, 9.5–9.8. Iconic lifting steel-frame bridge clad in a medieval-style stone cloak.
**Millennium Bridge** (TQ 320/806): see Fig 8.6. Incredibly level suspension bridge, a notable structure despite its teething problems.
**Hungerford Bridge** (TQ 305/803): railway bridge built on the piers from Brunel's Hungerford Suspension bridge.
**Albert Bridge, Chelsea** (TQ 274/775): see Fig 8.26. Very ornate early suspension and cable stay structure.
**Wharncliffe Viaduct, Hanwell:** (TQ 150/804) see Fig 10.8. Brunel's elliptical arch brick viaduct.

## GREATER MANCHESTER
**Marple Viaduct, Marple** (SJ 955/901): see Fig 6.8. Girder bridge. Fig 10.11. Stone viaduct and bridge running parallel with the canal aqueduct.
**Marple Aqueduct** (SJ 955/900) see Fig 11.9. Huge stone aqueduct with relieving holes in the spandrels.

FIG. 12.7: SALFORD QUAYS, GREATER MANCHESTER

**Barton Swing Aqueduct** (SJ 767/976): see Figs 9.13, 9.14. Unique major swing aqueduct, adjacent to road swing bridge (Fig 9.12). Arch from earlier aqueduct just north of it (Fig 11.3).
**Salford Quays, Greater Manchester** (SJ 830/975): see Fig 12.7.
**Castlefields Railway Bridge** (SJ 833/975): see Fig 5.13. Ornate cast-iron railway bridge with Gothic decoration.
**Stockport Viaduct** (SJ 890/904): Dramatic brick viaduct which now crosses the M60 near junction 1–2.
**Warburton Bridge** (SJ 696/902) see Fig 7.5. Cantilever bridge across Manchester Ship Canal.
**Cable Stayed Footbridge, M62 Junc 6–7** (SJ 801/927) see Fig 8.28.

## NORTHAMPTONSHIRE
**Cosgrove Canal Bridge** (SP 793/426) see Fig 4.10. Interesting Gothic-style canal bridge.

**Duddington Bridge** (SK 986/009) see Fig 2.25. Medieval bridge in idyllic stone village.
**Welland Viaduct, Harringworth** (SP 913/975) see Figs 10.6, 10.10. Longest viaduct outside London. Huge!

## NORTHUMBERLAND
**Willowford Roman Bridge, Hadrian's Wall** (NY 622/664) see Figs 1.4, 2.27. Remains of the eastern abutment and part of the wall.
**Chesters Roman Bridge, Hadrian's Wall** (NY 914/701) see Figs 1.13, 2.26. Remains of the eastern abutment and part of the wall.
**Corbridge** (NY 989/642) see Fig 3.15. Imposing large stone bridge from 1674.

## OXFORDSHIRE
**Wallingford Bridge** (SU 610/894)
**Abingdon Bridge** (SU 499/968) Much restored medieval bridge over the Thames; see Fig 5.3. Small cast-iron bridge over the Ock.
**Culham Bridge** (SU 501/957) see Fig 2.39. Small 15th-century bridge.
**Blenheim Palace Bridge** (SP 439/164) see Fig 4.11. Large single arch designed by Vanburgh.

## SHROPSHIRE
**Ironbridge** (SJ 672/034) see Figs 5.1, 5.2. Famous original iron-arch bridge over the River Severn.
**Cable Stayed Footbridge, Shrewsbury** (SJ 491/128) see Fig 8.28.
**Welsh Bridge, Shrewsbury** (SJ 489/127) see Fig 4.5. Classical style 18th-century bridge.
**English Bridge, Shrewsbury** (SJ 496/124) Ornate classical bridge from 1769.

**Atcham Bridge** (SJ 540/093) see Fig 4.21. Stunning classical bridge across the Severn.
**Atcham New Bridge** (SJ 540/093) see Fig 5.23. Concrete replacement for old bridge but with decoration to complement it.
**Kingsland Toll Bridge, Shrewsbury** (SJ 488/121) see Fig 5.19. Metal arch with a suspended roadway.
**Coalport Bridge** (SJ 702/021) see Fig 5.10. Impressive cast-iron bridge down-stream of Ironbridge.
**Cantlop Bridge** (SJ 517/062) see Fig 5.11. Small cast-iron road bridge by Telford.
**Albert Edward Bridge, Ironbridge** (SJ 660/038) see Fig 5.18. Huge cast-iron arch over the River Severn by John Fowler.
**Jackfield Bridge, Ironbridge** (SJ 681/033) see Fig 8.25. New cable stay bridge.
**Castle Bridge, Shrewsbury** (SJ 498/130) see Fig 7.6. Concrete cantilever footbridge.
**Porthill Suspension Bridge, Shrewsbury** (SJ 485/126) see Fig 8.15.
**Coalbrookdale Viaduct** (SJ 668/049) see Fig 10.3.
**Longton on Tern Aqueduct** (SJ 617/156) see Fig 11.2. One of the first iron aqueducts.

## STAFFORDSHIRE
**Three Shires Head, nr Leek** (SK 009/685) see Fig 1.14.
**Dove Bridge, Uttoxeter** (SK 105/344 ) see Fig 3.7. 14th-century with later rounded arches.
**Dove Suspension Bridge, Uttoxeter** (SK111/339 ) see Fig 8.15.

## Bridges to Visit

**Essex Bridge, Shugborough** (SJ 995/226) see Fig 3.9. Long and narrow multi arched bridge from the 16th century.
**Ilam Bridge** (SK 135/ 508) see Fig 4.15. 19th-century Gothic-style bridge.
**Denford** (SJ 959/ 536) see Fig 4.26. One of a number of simple, stone bridges along the Caldon Canal.
**Stretton Aqueduct** (SJ 873/107) see Fig 11.14. Black and white iron trough across the A5.

### SURREY
**Tilford Bridge** (SU 873/435) see Fig 2.14. 13th-century bridge with wooden handrail.

### SUSSEX (WEST)
**Stopham Bridge** (TQ 029/184) see Fig 3.12. Attractive sandstone bridge with a later central arch.
**Ouse Valley Viaduct, Balcombe** (TQ 323/278) see Fig 10.9. Notable early brick viaduct.

### TEESIDE
**Middlesbrough Transporter Bridge** (NZ 500/213) see Fig 9.17. Large cantilever transporter bridge.

### TYNESIDE
**Tyne Bridge, Newcastle** (NZ 245/638) see Figs 5.21, 5.22. Famous steel arch bridge with suspended roadway and Art Deco towers.
**A189 Road Bridge, Newcastle** (NZ 245/631) see Fig 6.13. Concrete road bridge over the Tyne.
**High Level Bridge, Newcastle** (NZ 251/637) see Fig 6.17. Stephenson's tied arch railway and road bridge across the Tyne.

FIG. 12.8: KILDWICK BRIDGE, YORKS

**Newcastle Swing Bridge** (NZ 252/637) see Figs 9.10, 9.11. Early major swing bridge with first use of hydraulic accumulator.
**Gateshead Millennium Bridge** (NZ 257/639) see Fig 9.18, 9.19. The famous blinking bridge over the Tyne.

### WORCESTERSHIRE
**Bewdley Bridge** (SO 787/754) see Fig 4.23. Fine classical bridge from 1799 by Telford.

### YORKSHIRE (SOUTH)
**Chantry Bridge, Wakefield** (SE 338/201) see Fig 2.19. 14th-century bridge with chapel.
**Chantry Bridge, Rotherham** (SK 427/930) see Fig 2.20. 15th-century bridge with chapel.
**Locomotive Bridge, Huddersfield** (SE 149/168) see Fig 9.4. Rare early horizontally lifting bridge.

# BRIDGES EXPLAINED: VIADUCTS & AQUEDUCTS

**FIG. 12.9: PIERCEBRIDGE, NORTH YORKSHIRE**

Lockwood Viaduct, Huddersfield (SE 133/147) see Fig 10.18.

### YORKSHIRE (WEST)
**Hebden Bridge** (SD 992/273) see Fig 3.8. 16th-century packhorse bridge in middle of town.
**Scammonden Bridge** (SE 046/168) see Fig 5.25. Huge concrete arch across the M62.

### YORKSHIRE (NORTH)
**Kildwick Bridge** (SE 011/457) see Fig 12.8. Stone arch bridge dating from 1305.
**Piercebridge Roman Bridge** (NZ 214/155) see Fig 6.6. Remains of an abutment and bridge base dating from 2nd century AD.
**Piercebridge** (NZ 211/156) see Fig 12.9. 18th-century stone arch bridge.
**Ribblehead Viaduct** (SD 759/794) see Fig 10.1. Huge stone viaduct across windswept moorland.

### WALES
**Menai Suspension Bridge, Bangor, Gwynedd** (SH 557/714) see Figs 1.10, 8.1, 8.7, 8.8. Telford's famous chain suspension bridge. Notable museum in the village hall on Anglesey side, well worth a visit.
**Britannia Bridge, Gwynedd** (SH 541/710) see Figs 6.9–6.11. Stephenson's famous box-girder bridge of which the towers survive; a section of the box still stands on the mainland side.
**Waterloo Bridge, Betws-y-Coed, Gwynedd** (SH 798/557) see Fig 5.16. Unique cast-iron bridge with an inscription along arch.
**Betws-y-Coed Suspension Bridge** (SH 796/565) see Fig 8.15.
**Pontcysyllte Aqueduct, Trevor, Clywd** (SJ 270/420) see Figs 1.15, 11.1, 11.12, 11.13. The most famous and spectacular British aqueduct by Telford.
**Cysylltau Bridge, Trevor** (SJ 268/420) see Fig 12.10. Medieval stone arch bridge
**Bangor Bridge, Bangor-is-y-Coed Bridge, Clwyd** (SJ 388/454) see Fig 3.13. Large sandstone bridge over River Dee.
**Monnow Bridge, Monmouth** (SO 504/124) see Fig 2.22: Unique 13th-century bridge with gatehouse.
**Llangollen Ancient Bridge, Denbighshire** (SJ 215/422) see Fig 2.28. 13th-century bridge, although restored, in a prominent position in the town.

# BRIDGES TO VISIT

**Pontypridd, Glamorgan** (ST 074/904) see Fig. 4.1. Massive, narrow single segmental arch with flood holes.
**Chepstow Bridge, Gwent** (ST 536/943) see Fig 5.17. Large cast-iron bridge by Rennie.
**Conwy Tubular Bridge** (SH 785/775) see Fig 6.12. The only surviving Stephenson box-girder bridge, runs parallel to the suspension bridge.
**Conwy Suspension Bridge** (SH 785/775) see Fig 8.9. Telford's smaller suspension bridge in original condition (National Trust: limited opening times)
**Wye Bridge, Gwent** (ST 543/911) see Fig 8.27. Continuation of the Severn Suspension Bridge. with early use of cable stays.
**Newport Transporter Bridge, Gwent** (ST 318/862) see Figs 9.15, 9.16. One of only two working transporter bridges in Britain.
**Chirk Viaduct, Clywd** (SJ 287/373) see Fig 10.14. Stone viaduct running alongside the aqueduct.
**Chirk Aqueduct, Clwyd** (SJ 287/373) see Figs 11.10, 11.11. Telford's stone and iron aqueduct.

## SCOTLAND
**Tay Bridge, Dundee** (NO 391/280) see Figs 1.1, 1.2. Replacement bridge for Bouch's ill-fated design.
**Forth Rail Bridge, Lothian** (NT 135/795) see Figs 7.1, 7.3, 7.4. Probably

FIG. 12.10: CYSYLLTAU BRIDGE, TREVOR

the most famous bridge in the country, with its distinctive red oxide coat
**Forth Road Bridge, Lothian** (NT 125/796) see Fig 8.16. First of the new generation of suspension bridges built after the Second World War.

NOTE: The bridges listed are not necessarily open to the public or safe to cross. They can all be viewed from close quarters from public footpaths. The best route to approach them can be seen on the relevant Ordnance Survey map.

# Glossary

**ABUTMENT:** A support on a bank at the side of a bridge built to resist the outward or vertical thrust of the structure

**AIR SPINNING:** The method of using spools of wire run back and forth to create a continuous cable on a suspension bridge

**ANCHORAGE:** A fixing in the ground on each side of a suspension bridge for the cables or chains

**AQUEDUCT:** A bridge or channel for carrying water on a level

**BASCULE:** A see-saw type of bridge with a counterweight on one end of a hinged deck

**BATTER:** The sloping face of a wall which is wider at the bottom and narrows towards the top, designed for stability

**BEAM:** A horizontal load-bearing structural element

**BEDROCK:** A solid layer of rock beneath the silt in a river bed

**BOWSTRING ARCH OR TIED ARCH:** A form of truss with an arch tied at each end by a horizontal element which counters its outward thrust (forms the shape of a bow from a bow and arrow weapon)

**BOX GIRDER:** A form of truss consisting of a hollow tubular box

**CABLE STAY BRIDGE:** A bridge in which the deck is directly supported by cables hanging down from the towers or pylons

**CAISSON:** A hollow metal or concrete circular vessel which is sunk vertically to the river bed to allow work at the foundations, part or all of which is usually left in place when the bridge is complete

**CANTILEVER:** A beam which is supported at one end by a fixing in a wall or on a diagonal arm

**CAST IRON:** A type of iron which is formed in moulds; it is stronger in compression

**CATENARY:** An inverted arc formed by a cable or chain hanging from two level points, for example, the towers in a suspension bridge

**CENTRING:** A temporary timber frame used to support an arch under construction

**CHORD:** A horizontal member of a truss

**COFFERDAM:** A temporary watertight enclosure which is pumped dry to enable work to be carried out below the water level

**COMPRESSION:** The pushing force which compacts a material inwards

**CORBEL:** A stone or wooden bracket used to support centring or for decorative effect

**CUTWATER:** A pointed or rounded face to a pier

**DEAD LOAD:** The weight of a structure alone

**ELLIPTICAL ARCH:** An arch with two smaller radii at each end of a larger one

# Glossary

| | |
|---|---|
| **GIRDER:** | A metal or concrete beam usually with an I-, U- or T-shaped profile |
| **HANGER:** | A vertical suspension cable or rod connecting the deck to the main cable or chain |
| **HAUNCH:** | The section of an arch between the springing and crown |
| **KEYSTONE:** | The voussoir at the crown of an arch |
| **LIVE LOAD:** | The weight of traffic acting upon the bridge |
| **PIER:** | A ground support between the abutments holding up the bridge, usually set in the river bed |
| **PLATE GIRDER:** | A horizontal truss made from a girder frame with sheet metal between, used for parapets or in series under the deck |
| **POINTED ARCH:** | An arch in which two arcs intercept at the crown |
| **SEGMENTAL ARCH:** | A shallow arch having the form of a segment of a circle |
| **SEMI-CIRCULAR ARCH:** | An arch having the form of a half circle |
| **SHEAR:** | A strain produced by the force across a beam which can cause it to fail, one side sliding down past another |
| **SIDE SPAN:** | The outer section between the tower and anchorage of a suspension bridge |
| **SOFFIT:** | The underside of an arch |
| **SPANDREL:** | The face of a bridge between the arch and the parapet |
| **SPRINGING:** | The point from which an arch rises on an abutment or pier |
| **STARLING:** | A boat-shaped foundation around a pier designed to protect it under water from scouring. |
| **TENSION:** | The stretching force which tears material apart |
| **TORSION:** | The strain upon a material from twisting |
| **TRUSS:** | A horizontal framework of beams or girders used to support a long deck |
| **SUSPENDER:** | See hanger |
| **VOUSSOIR:** | A wedge-shaped stone or brick forming part an arch |
| **WROUGHT IRON:** | A type of iron which is malleable and is stronger in tension |

# BIBLIOGRAPHY

The following books may be useful for further information:

Richards, J.M. *The National Trust Book of Bridges* (1984)
Ashley, Peter *Bridging the Gap* (2001)
Slack, Margaret *The Bridges of Lancashire and Yorkshire* (1986)
Taylor, Christopher *Roads and Tracks of Britain* (1994)
Rowlands, M.L.J. *Monnow Bridge and Gate* (1994)
Burn-Murdoch, Bob *St Ives Bridge and Chapel* (2001)
Richardson, John *The Annals of London* (2000)
Simco, Angela and Peter McKeague *Bridges of Bedfordshire* (1997)
Cook, Peter *Medieval Bridges* (1998)
Blackwall, Anthony *Historic Bridges of Shropshire* (1985)
Brown, David J. *Bridges, Three Thousand Years of Defying Nature* (2005)
Gordon, J.E. *Structures or Why Things Don't Fall Down* (1991)
De Mare, Eric *Bridges of Britain* (1975)
McIlwain, John *Clifton Suspension Bridge* (2000)
Pevsner, Nikolaus *The Buildings of England series* (county by county)
Kirby, Richard Shelton, Sidney Withington, Arthur Burr Darling, Frederick Gridley Kilgour, *Engineering in History* (1990)

A lot of information can be found on the Internet. The following sites may be useful for general engineering and historic facts or details of specific bridges:

**Specific Bridges**
www.oldlondonbridge.com
www.severnbridge.co.uk
www.forthbridges.org.uk
www.eriding.net/dandt/structures.shtml (Humber Bridge)
www.towerbridge.org.uk
www.dundeecity.gov.uk (Tay Bridge)
www.taybridgedisaster.co.uk

**General Information**
www.brantacan.co.uk
www.wikipedia.org
www.scienceandsociety.co.uk
www.bbc.co.uk
www.geocities.com
www.clevelandbridge.com
www.engineering-timelines.com
www.icomos.org/studies/bridges.htm

# INDEX

## A

Albert Edward Bridge, Salop: 69, 152
Armstrong, Sir William: 119, 121
Ashford in the Water (Sheepwash Bridge), Derbys: 44, 148
Ashness Bridge, Cumbria: 147, 148

## B

Bakewell Bridge, Derbys: 37, 148
Balcombe Viaduct: 133, 153
Bangor-is-y-Coed Bridge, Clwyd: 47, 154
Barton Swing Aqueduct: 122–124, 139, 151
Bath (Pulteney Bridge): 56, 146
Bewdley Bridge, Worcs: 59, 153
Bickleigh Viaduct: 134, 149
Bideford Bridge, Devon: 38, 149
Bouch, Thomas: 7–8, 10, 12, 19, 84, 155
Brindley, James: 123, 138, 139, 140, 149
Britannia Bridge, Gwynedd: 77–79, 81, 82, 154
Bromham Bridge, Beds: 38, 146
Brunel, Isambard Kingdom: 15, 60, 72–73, 77, 81–82, 90, 98–100, 130, 133, 134, 146, 148, 149

## C

Cambridge, (Clare Bridge): 54, 147
Cannington Viaduct: 134, 149
Cantlop Bridge, Salop: 67, 152
Chapel-en-le-Frith viaducts: 131, 135, 148
Chepstow Bridge, Gwent: 69, 155
Chesters Bridge: 18, 35, 47, 152
Clifton Suspension Bridge: 98–101, 146
Coalport Bridge, Salop: 67, 152
Conway Suspension Bridge: 96–98, 155
Conwy Tubular Bridge, Gwynedd: 78, 155
Crowland Bridge, Lincs: 31, 151
Culham Bridge, Oxon: 38, 152

## D

Darby III, Abraham: 62–63
Digswell Viaduct: 135, 150
Dove Aqueduct: 140, 149
Duddington Bridge, Northants: 35, 152
Dundas Aqueduct: 140 146
Durham (Old Elvet Bridge): 37, 149

## E

Eamont Bridge, Penrith: 146, 148
Edwards, William: 48–49, 53, 142

## F

Forth Rail Bridge: 17, 65, 83, 84–87, 120, 127, 155
Forth Road Bridge: 92, 101–103, 155
Fowler, John: 69, 84, 152

## G

Gateshead Millennium Bridge: 127–128, 153
Great Barford Bridge, Beds: 49, 146
Gwynn, John: 57, 58, 71

## H

Harrison, Thomas: 57, 58
Harrold Bridge, Beds: 36, 146
Hereford (Wye Bridge): 19, 150
Hebden Bridge, Yorks: 45, 154
Hinchingbrooke (Nun's Bridge), Huntingdon: 37, 147
Holt Bridge, Cheshire: 34, 36, 147
Humber Bridge: 104–106, 151
Hungerford Bridge, London: 99, 151
Huntingdon Bridge: 29, 31, 147

## I

Ironbridge: 10, 61–63, 152

## J

Jackfield Bridge, Salop: 107, 152

## K

Kildwick Bridge, North Yorks: 153, 154
Kingsferry Lift Bridge: 116, 150
Kirkby Lonsdale (Devil's Bridge), Cumbria: 41, 148

# INDEX

## L

Lanercost Bridge, Cumbria: 46, 148
Lincoln High Bridge: 34, 151
Llangollen Bridge: 36, 154
London Bridge: 21, 22, 34, 40–41
Longdon on Tern Aqueduct: 138, 141–144, 152
Lune Aqueduct: 139–142, 150

## M

Maidenhead Road Bridge, Berks: 57, 146
Maidenhead Rail Bridge, Berks: 60, 146
Marlow Suspension Bridge: 97, 146
Medbourne Bridge, Leics: 30, 150
Meldon Viaduct: 130, 132, 149
Menai Straits Suspension Bridge: 16, 89, 92, 94–97, 154
Middlesbrough Transporter Bridge: 126–127, 153
Millennium Footbridge, London: 92, 93, 151
Monnow Bridge, Monmouth: 33–34, 154

## N

Newport Transporter Bridge: 125–126, 155

## O

Oakley Bridge, Beds: 11, 146
Over Bridge, Glos: 58, 149

## P

Piercebridge, North Yorks: 75, 154
Pontsycyllte Aqueduct: 20, 137, 141–145, 154
Pontypridd Bridge, Glamorgan: 48, 51, 53, 155
Postbridge Clapper Bridge, Devon: 74, 149
Preston Bridges: 130, 150

## Q

Queen Elizabeth II Bridge: 111–112, 150

## R

Rennie, John: 69, 95, 139–141, 155
Ribblehead Viaduct: 129, 131–132, 154
Rotherham Chantry Bridge: 32–33, 153
Royal Albert Bridge, Saltash: 72–73, 81–82, 148
Runcorn (Silver Jubilee Bridge), Cheshire: 70, 147
Runcorn (Railway Bridge), Cheshire: 80, 147

## S

St Ives Bridge, Cambs: 23, 32, 33, 147
Scammonden Bridge, West Yorks: 71, 154
Second Severn Crossing: 109, 112–113, 149
Severn Suspension Bridge (First): 103–104, 109, 149
Shugborough (Essex Bridge), Staffs: 45, 153
Sonning Bridge, Berks: 40, 146
Stephenson, Robert: 76–79, 80, 81, 95, 100, 153, 155
Stockport Viaduct: 130, 151
Stony Stratford (Ouse Bridge), Bucks: 57, 147
Stopham Bridge, Sussex: 46, 153
Stretton Aqueduct: 145, 153
Sutton Bridge, Beds: 39, 146

## T

Tamar Road Bridge: 102, 148
Tay Rail Bridge: 7–8, 12, 13, 14, 18–19, 84, 155
Telford, Thomas: 16, 50, 53, 58, 59, 67, 68, 78, 90, 92, 94–98, 137, 138, 139, 141–145, 147, 149, 152–155
Thelwall Viaduct: 136, 147
Thornborough Bridge, Bucks: 31, 36, 46, 146
Tickford Bridge, Bucks: 67, 147
Tilford Bridge, Surrey: 30, 153
Tower Bridge: 114, 116–120, 151

## U

Uttoxeter (Dove Bridge), Staffs: 44, 100, 152

## W

Wakefield Chantry Bridge: 32–33, 153
Warburton Bridge, Cheshire: 87, 151
Welland Valley Viaduct: 131, 132, 133, 152
Wharncliffe Viaduct: 133, 151
Willowford Bridge, Northumberland: 10, 35, 152
Wilton Bridge, Ross on Wye: 47, 150
Wye Bridge, Gwent: 109, 155